Measuring Convergence in Science and Engineering

PROCEEDINGS OF A WORKSHOP

Shadya Sanders, *Rapporteur*

Committee on National Statistics

Division of Behavioral and Social Sciences and Education

The National Academies of
SCIENCES · ENGINEERING · MEDICINE

THE NATIONAL ACADEMIES PRESS
Washington, DC
www.nap.edu

THE NATIONAL ACADEMIES PRESS 500 Fifth Street, NW Washington, DC 20001

This activity was supported by a contract between the National Academy of Sciences and the National Science Foundation (award #1822391). Any opinions, findings, conclusions, or recommendations expressed in this publication do not necessarily reflect the views of any organization or agency that provided support for the project.

International Standard Book Number-13: 978-0-309-22183-2
International Standard Book Number-10: 0-309-22183-8
Digital Object Identifier: https://doi.org/10.17226/26040

Additional copies of this publication are available from the National Academies Press, 500 Fifth Street, NW, Keck 360, Washington, DC 20001; (800) 624-6242 or (202) 334-3313; http://www.nap.edu.

Printed in the United States of America

Suggested citation: National Academies of Sciences, Engineering, and Medicine. (2021). *Measuring Convergence in Science and Engineering: Proceedings of a Workshop.* Washington, DC: The National Academies Press. https://doi.org/10.17226/26040.

The National Academies of
SCIENCES • ENGINEERING • MEDICINE

The **National Academy of Sciences** was established in 1863 by an Act of Congress, signed by President Lincoln, as a private, nongovernmental institution to advise the nation on issues related to science and technology. Members are elected by their peers for outstanding contributions to research. Dr. Marcia McNutt is president.

The **National Academy of Engineering** was established in 1964 under the charter of the National Academy of Sciences to bring the practices of engineering to advising the nation. Members are elected by their peers for extraordinary contributions to engineering. Dr. John L. Anderson is president.

The **National Academy of Medicine** (formerly the Institute of Medicine) was established in 1970 under the charter of the National Academy of Sciences to advise the nation on medical and health issues. Members are elected by their peers for distinguished contributions to medicine and health. Dr. Victor J. Dzau is president.

The three Academies work together as the **National Academies of Sciences, Engineering, and Medicine** to provide independent, objective analysis and advice to the nation and conduct other activities to solve complex problems and inform public policy decisions. The National Academies also encourage education and research, recognize outstanding contributions to knowledge, and increase public understanding in matters of science, engineering, and medicine.

Learn more about the National Academies of Sciences, Engineering, and Medicine at **www.nationalacademies.org**.

The National Academies of

SCIENCES · ENGINEERING · MEDICINE

STEERING COMMITTEE FOR THE WORKSHOP ON THE IMPLICATIONS OF CONVERGENCE FOR HOW THE NATIONAL CENTER FOR SCIENCE AND ENGINEERING STATISTICS MEASURES THE SCIENCE AND ENGINEERING WORKFORCE

BARBARA ENTWISLE (*Chair*), Department of Sociology, University of North Carolina at Chapel Hill
KARA HALL, National Cancer Institute, National Institutes of Health
JAMES BRITT HOLBROOK, Department of Humanities, New Jersey Institute of Technology
PRAMOD P. KHARGONEKAR, School of Engineering, University of California, Irvine
JULIE THOMPSON KLEIN, English Department, Wayne State University (emerita) and Transdisciplinarity Lab, ETH Zurich
JASON OWEN-SMITH, Institute for Social Research, University of Michigan
JOLENE SMYTH, Department of Sociology, University of Nebraska

KRISZTINA MARTON, *Study Director*
ELLIS GRIMES, *Senior Program Assistant*
TINA LATIMER, *Senior Program Assistant*
BRIAN A. HARRIS-KOJETIN, *Director, Committee on National Statistics*

COMMITTEE ON NATIONAL STATISTICS

Acknowledgments

This Proceedings of a Workshop was reviewed in draft form by individuals chosen for their diverse perspectives and technical expertise. The purpose of this independent review is to provide candid and critical comments that will assist the National Academies of Sciences, Engineering, and Medicine in making each published proceedings as sound as possible and to ensure that it meets the institutional standards for quality, objectivity, evidence, and responsiveness to the charge. The review comments and draft manuscript remain confidential to protect the integrity of the process.

We thank the following individuals for their review of this proceedings: Steve W.J. Kozlowski, Department of Psychology, University of South Florida, and Nicole Motzer, National Socio-Environmental Synthesis Center, University of Maryland.

Although the reviewers listed above provided many constructive comments and suggestions, they were neither asked to endorse the content of the proceedings nor allowed to see the final draft before its release. The review of this proceedings document was overseen by Jennifer S. Bond, Senior Fellow, Council on Competitiveness. She was responsible for making certain that an independent examination of this proceedings document was carried out in accordance with the standards of the National Academies and that all review comments were carefully considered. The views expressed in this document reflect both individual and collective opinions of the meeting participants and not necessarily those of the National Academies.

This Proceedings of a Workshop was drafted by Shadya Sanders, Dana Glei, Rose Li, Dana Carluccio, and Nancy Tuvesson of Rose Li & Associates, Inc. Responsibility for the final content rests entirely with the rapporteurs and the National Academies.

Contents

Appendixes

1

Introduction

BACKGROUND

This Proceedings of a Workshop summarizes the presentations and discussions at the Workshop on the Implications of Convergence for How the National Center for Science and Engineering Statistics (NCSES) Measures the Science and Engineering Workforce, which was held virtually and livestreamed on October 22–23, 2020. The workshop was convened by the Committee on National Statistics (CNSTAT) to help NCSES, a division of the National Science Foundation (NSF), set an agenda to inform its methodological research and better measure and assess the implications of convergence for the science and engineering (S&E) workforce and enterprise. While various definitions exist, the NSF website describes convergence research as having two primary characteristics: It is driven by a specific and compelling problem, and it requires deep integration across disciplines. The convergence paradigm "intentionally brings together intellectually diverse researchers to develop effective ways of communicating across disciplines by adopting common frameworks and a new scientific language. This may in turn help solve the problem that necessitated the collaboration, resulting in novel ways of framing research questions and opening new research vistas."[1]

The workshop was structured to bring together scientists and researchers from multiple disciplines, along with experts in science policy, university administration, and other stakeholders to review and provide input on

[1]See Convergence Research at NSF at https://www.nsf.gov/od/oia/convergence/index.jsp.

defining and measuring convergence and its impact on science and scientists. The workshop agenda is provided in Appendix A. Biographical sketches of the steering committee members and workshop presenters are included in Appendix B.

CNSTAT has a long history of providing advice on matters of science to the federal government and conducting independent reviews of federal statistical activities to improve the quality of statistical information, contribute to statistical policies, and coordinate the federal government's activities aimed at providing a forward-looking vision for the federal statistical system. CNSTAT's director, Brian Harris-Kojetin, described measurement as a frequent focus of CNSTAT's work and as being fundamental to efforts by statistical agencies to provide reliable statistics on various topics. It requires statistical agencies to develop a deep understanding of their topic areas and clear definitions for what should be measured. This knowledge includes understanding what information exists, whether this information is well organized, how it can be accessed, and how agencies interpret the data, particularly if they are asked to provide information about them. Statistical agencies excel at the details that must be considered to accurately measure the nuances of a particular topic. The challenge to accurate measurement is that our society and economy are constantly evolving, requiring innovation in methods of measurement. Harris-Kojetin thanked NCSES for the opportunity to convene this workshop.

WORKSHOP FOCUS

The purpose of the workshop was to help NSF and NCSES "set an agenda to inform its methodological research and improve its data collection programs to better measure and assess the implications of convergence for the S&E workforce and enterprise." Entwisle explained how convergence has expanded beyond the term coined nearly 10 years ago when MIT professors Phil Sharp and Robert Langer published their work on convergence and biomedical science in 2011.[2] Initially, convergence focused on joining engineering, physical science, and life science within academic institutions. It now has expanded to encompass novel disciplines with new outcomes. The workshop sought to explore the status of the definitions of convergence, its practice, its outputs, its effects, and promising paths forward for NCSES in this area. A central objective of the workshop was to help NCSES refine and prioritize its data collection and methodological research on convergence. Box 1-1 shows the workshop's full statement of task.

---

[2]Sharp, P.A., and R. Langer. (2011). Research agenda. Promoting convergence in biomedical science. *Science* 333(6042):527. doi:10.1126/science.1205008.

BOX 1-1
Statement of Task

The Committee on National Statistics of the National Academies of Sciences, Engineering, and Medicine proposes to hold a 2-day open workshop to help the National Science Foundation's National Center for Science and Engineering Statistics (NCSES) set an agenda to inform its methodological research and improve its data collection programs to better measure and assess the implications of convergence for the science and engineering workforce and enterprise. A planning committee will organize the workshop, which will bring together scientists from a variety of disciplines as well as experts in science policy, university administrators, and other stakeholders to review and provide input on defining convergence and measuring its impact on science and scientists. The resulting workshop will be oriented to help NCSES refine and prioritize its data collection and methodological research in this area. Proceedings of the Workshop will be prepared by a designated rapporteur in accordance with institutional guidelines.

THE NATIONAL SCIENCE FOUNDATION'S INTEREST IN MEASURING CONVERGENCE

Arthur (Skip) Lupia (assistant director, NSF) described 2020 as an ideal time to consider convergence because of the enormous challenges facing not only the United States but also countries across the globe. The COVID-19 pandemic has impacted lives everywhere, for example, in terms of health, education, business, and social supports. As one way of improving the quality of life in these circumstances, Lupia called attention to NSF's Convergence Accelerator Program, which accelerates use-inspired convergence research in areas of national importance.[3]

Quoting American chemist Linus Pauling, he proposed "the best way to have good ideas is to have a lot of ideas." Lupia added that the types of good ideas that can have the largest social impact are almost always novel combinations of existing concepts. The challenge for an agency such as NSF is that many "novel combinations" fall between the cracks of existing disciplinary categories. This is one reason that NSF seeks to develop new programs like the Convergence Accelerator.

Lupia identified interdisciplinarity, multidisciplinarity, and convergence as concepts that can energize and leverage ideas to effect new ways of thinking. These terms are often used interchangeably, but they have different

[3]National Science Foundation. (October 9, 2020). Dear Colleague Letter: Request for Information on Future Topics for the NSF Convergence Accelerator (NSF 21-012). Available: https://www.nsf.gov/pubs/2021/nsf21012/nsf21012.jsp.

meanings (their definitions will be discussed throughout the workshop). Understanding that workshop participants would likely disagree about the meanings of these terms and the best ways to describe them, he tried to set expectations by providing an example of challenges associated with defining the term "implicit bias." He and Jon Krosnick convened a workshop of leading experts on this topic at NSF. During that workshop, the experts spent 2 days engaged in rigorous and illuminating discussions. On day 3, they asked the experts to define implicit bias. It took more than 2 hours and was much more difficult than anyone expected.

Lupia then turned to the topic of measurement. NSF's Social, Behavioral and Economic Sciences Directorate (SBE) seeks to support "rigorous, precise research that empowers people and improves quality of life." SBE does this by supporting groundbreaking research, the creation of fundamental data and analytic infrastructure, and essential training. He also outlined SBE principles of rigor (the ability to explain how we know what we know), ethics (empowering people through awareness of the whole scientific workflow), precision in measurement and conceptualization, and causality.

NCSES is part of SBE and is tasked with measuring key aspects of the nation's science and engineering ecosystem. It must deliver measures that meet SBE and other federal standards. Moreover, the value of NCSES measures depends on their being broadly viewed as valid and legitimate by the research community.

With these criteria in mind, Lupia indicated that the concept of convergence is potentially ambiguous, and wondered whether it might be better to treat it as an aspirational idea rather than something to measure explicitly. To assess this possibility, Lupia posed five critical questions for the group:

1. Is a valid and broadly legitimate convergence measure possible, that is, a measure that signals standardization and can be reliably and consistently derived from the activities of diverse institutions?
2. Is it possible to create mutual exclusivity and collective exhaustion in critical aspects of the underlying concept? Is neuroscience, for instance, an example of convergent work, or is it its own discipline?
3. If we can answer the first two questions, who does the measure empower to effectively accomplish high-value goals to a greater degree?
4. If we answer all the preceding questions, what opportunity costs exist for NCSES to develop such a measure?
5. Having answered these questions, is convergence more effective as a motivational ideal than a measured phenomenon?

After Lupia's presentation, Khargonekar asked whether Lupia is considering single or multiple measures of convergence when posing these questions. Lupia responded that his view of convergence is a highly dimensional concept, and if other questions about the measure can be answered, he has no preconceived notion as to whether convergence should have a single measure or multiple measures. Holbrook followed by asking whether it is possible to answer both the first and fifth questions in the affirmative. Lupia expressed his belief that it is logically possible to answer both questions in the affirmative.

WORKSHOP GOALS FOR THE NATIONAL CENTER FOR SCIENCE AND ENGINEERING STATISTICS

Emilda Rivers (director, NCSES) described the needs, goals, and challenges for NCSES relative to the measurement of convergence. She agreed with Lupia that one or several concepts could be related to convergence and that NCSES' goal would be to convey the reliability, precision, and transparency of the measurement process. This goal could become increasingly difficult to achieve if individual communities hold different definitions of convergence. Rivers organized her presentation as a set of questions directed to the steering committee. First, she asked whether any definitional issues could be overcome. Goals are sometimes framed with a policy emphasis that may differ significantly from an emphasis on measurement, and this difference can result in the loss of nuance and lead to more confusion rather than clarity. She then inquired about (1) the unintended consequences of measuring convergence and the gaps that may exist when providing a framework at the national level, (2) the communication of gaps where they exist, and (3) the means for accurately reflecting the impact of convergence on the scientific community.

As a federal agency, NCSES is obligated to provide information that is influential across the nation. NCSES requires high-quality data and must be able to communicate both the level of expertise involved in the production of data used and the impact of the data to people unfamiliar with convergence. After providing these questions to the committee, Rivers wondered about the timing of this measurement goal. As Entwisle mentioned, it may be too early to try to derive a national measure, given that the concept of convergence was originally defined only a little more than a decade ago.

Erin Leahey (University of Arizona) asked whether a national measure is one that seeks consistency on a national scale (or in other words, uses the same measure used throughout the United States) or an international scale (that is, a measure that allows for comparison across different countries). Rivers responded that a national measure would consist of data that are representative of convergence across the United States. She expressed a reluctance toward thinking too broadly as well as a hope in the adoption of a national estimate of convergence across the United States as an initial goal. Bethany Laursen (Michigan State University) asked whether NCSES is willing to consider qualitative as well as quantitative measures of convergence. Rivers responded that NCSES plans to focus on a quantitative measure at this time. She clarified that multiple phases of qualitative research to improve understanding often occur before a quantitative measure is reached. Qualitative research can thus play a very important role.

ORGANIZATION OF THE REPORT

This summary describes the workshop presentations and discussions that followed each topic. The chapters are organized around the sessions of the workshop, which included overviews of the prior work of NCSES in this area and of the definitions of convergence, discussions of how convergence is practiced in a variety of settings, the outcomes and impacts of convergence, and existing efforts to measure convergence using a variety of approaches. The final sessions of the workshop focused on questions related to the feasibility of collecting national-level data on convergence, implementation considerations, and discussions among workshop participants about how NCSES could move the work in this area forward.

The full meeting agenda and biographical sketches of steering committee members and workshop presenters appear in the appendixes. This proceedings has been prepared by the workshop rapporteur as a factual summary of what occurred at the workshop. The planning committee's role was limited to planning and convening the workshop. The views contained in the proceedings are those of individual workshop participants and do not necessarily represent the views of all workshop participants, the planning committee, or the National Academies of Sciences, Engineering, and Medicine.

2

Setting the Stage: Why Is Convergence Important?

Joseph DeSimone (Stanford University and Carbon, Inc.) began with a story about setting and upholding one's stated goals. DeSimone recently restarted his academic lab and opened his first research group meeting by clearly displaying his group's values. These values included commitments to multiple dimensions of diversity, which DeSimone considers a fundamental tenet of innovation. He explained that when you are clear about these values, you become a destination for people seeking to achieve excellence. By contrast, remaining silent about these values can quickly cause problems. DeSimone presented research as an expression of faith about the possibility of innovation, the possibility that "new things can be discovered, [and] that newer can be better."[1] Academic research and, in DeSimone's view, research-driven startups both constitute "a form of optimism about the human condition."

Innovation was categorized by the economist Clay Christensen into three types: disruptive, sustaining, and efficiency. DeSimone explained that disruptive innovations transform the complex and expensive to make them affordable and accessible. A long-term investment, for example, may create completely new categories of science and sectors of industry such as the development of DNA sequencing and its evolution into the biotechnology industry. Disruptive innovations thus create millions of jobs. On the other hand, sustaining innovations do not create jobs, but they do preserve them by replacing old products with newly created components that improve

---

[1]Rosovsky, H. (1991). *The University: An Owner's Manual.* New York: W.W. Norton and Company.

product quality and support a healthy economy. Efficiency innovations streamline processes and reduce costs, but they eliminate jobs. In the automotive industry, for example, automation and technology innovations have eliminated portions of the workforce. Considering each aspect of efficiency, DeSimone asked the workshop participants where they believe they stand on the spectrum of innovation types and where they would like to be.

DeSimone argued that innovation can be driven by convergence. He described a framework by Susan Hockfield and her team at MIT that sees the convergence of the life sciences, physical sciences, and engineering as a strategy for driving innovation. DeSimone explained that an expansion toward a broader inclusivity of the disciplines (including the social sciences, humanities, and performing arts) could also be a blueprint for innovation. He offered the example of designing earthquake-proof buildings in Haiti. To complete such work, one would need to understand not only engineering but also French culture, language, history, and government policy. DeSimone further cited Apple and its former CEO Steve Jobs as an example of the inadequacy of any single field (technology, in this case). Jobs stated, "It's technology married with liberal arts married with the humanities that yields us the result that makes our heart sing." DeSimone also warned that technology companies at times lack a liberal arts perspective. Facebook, for example, was intended to bring people together but has incited division, and Juul set out to end smoking, but its replacement—vaping—instead became a serious problem. DeSimone argued that perspectives from both the humanities and social sciences should be more deeply embedded in the goals and the operation of technological initiatives.

Convergence also impacts professional training. DeSimone used the model of I-, T-, and π-shaped people to exemplify this impact. I-shaped individuals have strong training in their core fields, T-shaped people are deeply knowledgeable about a particular subject but can also collaborate well with people in other fields, and π-shaped individuals are deeply knowledgeable about multiple subjects. This typology can also be described in terms of language: T-shaped individuals may work across distinct disciplines that share a common language, whereas those who are π-shaped are truly multilingual with a deep understanding of multiple subjects that use different conceptual vocabularies. While DeSimone views π-shaped knowledge as a higher calling, he expressed concern about the demands on people who develop multiple deep knowledge bases across their careers, and the extensive time that this process requires. Those in academic areas of research are thus significantly older than those in entrepreneurial spaces like Silicon Valley tech companies: the average age of a Research Project R01 grantee is around 40 years, compared to the average age of recipients of seed grants, who tend to be in their mid-twenties.

Beyond training and educational demands on multidisciplinary researchers, convergence also requires a level of vulnerability. To be an effective collaborator, you must be willing to clearly articulate the limits of your expertise—what you do not know as well as what you do—yet it is not common for organizations to reward scientists for being clear about what they do not know. It is important to create an environment that fosters this kind of clarity and allows for such learning without penalty.

DeSimone argued that social and intellectual diversity both constitute critical components of convergence and innovation. He began by reflecting on an invitation he received early in his career to join an innovation committee for a large European company. Upon first entering the company boardroom, he found that the committee was homogenous, composed entirely of white men who had all been trained in one of just two research groups. Despite his surface similarity to these members, DeSimone did not fit in and was not treated as a true part of the group. Instead, he was consistently cast as an outsider. In DeSimone's view, the group's lack of inclusivity put it at a structural disadvantage that hindered the fostering of the innovation for which it was formed.

This experience influenced DeSimone's public commitment to inclusivity in his home institution and research group. He described a later invitation he received to deliver a keynote address to the American Chemical Society at a time when the NAACP was calling for a boycott of the state of South Carolina in response to its display of the Confederate flag. DeSimone's research group decided to honor the boycott and withdraw from the conference. The bold stance to remain consistent with the group's core value of inclusivity had consequences: the conference was relocated,[2] which attracted the attention of national news organizations covering the story. More recently, a personal essay in the prestigious journal *Angewandte Chemie, International Edition* expressed frustration with the professional focus on increasing diversity. The now deleted essay sparked a mass resignation by 16 of the 44 international advisory board members, including DeSimone, and has since rallied public support for the importance of building a culture of inclusivity.

DeSimone suggested that diversity promotes innovation when people of different backgrounds come together to learn from each other (in other words, when they participate in a form of cultural convergence). DeSimone stated, "We learn the most from those we have the least in common with," both in disciplinary and cultural terms. DeSimone went on to describe the work of political scientist Scott Page about the power of diversity in

[2]Brennan, M. (2000). Polymer workshop defers to NAACP boycott of South Carolina. *Chemical & Engineering News Archive* 78(10):58. doi:10.1021/cen-v078n010.p058.

problem solving. In his book *The Difference*,[3] he shows that diversity does not always yield better results than ability, but it does so far more often than expected. DeSimone sought to quell the fears of those who may worry that diversity and inclusivity imply an end to meritocracy, clarifying that both ability and diversity matter. Teams require people who can successfully and skillfully play different functional roles. Team success also requires different viewpoints, cultural backgrounds, ethnicities, and genders.

DeSimone reminded the group that actualizing these important considerations requires resources, or as he put it, "a vision without resources is a hallucination." He pointed to important differences in how resources are allocated in research and business settings by using Peter Thiel's differentiation of "zero to one" ideas (or "rebel ideas" about which very few people agree) and "one to *n*" ideas (or ideas that build on those rebel ideas). DeSimone suggested that "zero to one" ideas face a difficult time securing traditional research funding, because an idea that few people agree with will not earn the broad support required for a traditional grant. An application submitted to the National Institutes of Health (NIH) must receive an "excellent" score from 10 reviewers to receive funding—if nine reviewers give it an "excellent" score and one scores it "very good," it will not be funded—whereas in the venture capital world, one might receive nine "no" scores and only a single "yes," but that is sufficient to secure an investment in a "zero to one" idea. De Simone emphasized the importance of "zero to one" ideas and of putting those ideas into action through translational research. In the words of Johann Wolfgang von Goethe, "Knowing is not enough: we must apply. Willing is not enough: we must do."

DeSimone concluded by summarizing the key ingredients of innovation, which include diverse design teams, a commitment to mentorships and apprenticeships, efforts to behave differently than those in the majority of a field, a pursuit of connections between fields while building partnerships with domain experts, working performed on a massive scale rather than a merely incremental one, and engagement with others. These ingredients demonstrate the blend of convergence and diversity, which are interrelated and critically important.

DISCUSSION

Holbrook suggested that if diversity is necessary for convergence, then standardizing the meaning of convergence might end up interfering with convergence. DeSimone agreed with this characterization. He suggested that requests for applications may enforce standardization and expressed

[3]Page, S.E. (2007). *The Difference: How the Power of Diversity Creates Better Groups, Firms, Schools, and Societies* (2007). Princeton, NJ: Princeton University Press. doi:10.2307/j.ctt7sp9c.

hope that they might become less stringent and more interesting by allowing researchers to follow their own research interests.

Entwisle asked how DeSimone would define a field and thus measure when researchers are bridging fields. DeSimone suggested that fields are a lingering product of the 18th-century organizational structures still embedded in modern universities. He reiterated his views about I-, T-, and π-shaped individuals and the importance of promoting deep learning across multiple disciplines. DeSimone himself was warned early in his career to be careful of engaging in too many collaborative efforts because they may harm his potential for promotion. He personally received support from his department chair and suggested that all researchers should be able to receive funding as independent investigators early in their careers to promote the autonomy and latitude to think broadly.

Entwisle noted that the application of research for developing startups has been central to DeSimone's career and thus asked whether he considers this work to be necessarily linked to convergence as a concept. DeSimone responded that it is not a necessary part of convergence but can be a defining element, in part because translational work is deeply inspiring. He further explained that startup development does not lend itself to every discipline, thus he would neither force that model onto a researcher nor negate those researchers who do link their work to business applications.

Peter Schiffer (Yale University) asked DeSimone to comment on the value of measuring the impacts and outcomes of past, present, and proposed convergent research. Entwisle suggested that this question probes whether researchers need to know in advance if research is convergent or transformational and whether convergence is a property of research itself. DeSimone said that convergence is valuable for social scientists to quantify but cautioned against prescribing how individual researchers should carry out their research. He argued that the national portfolio of research should be diverse. Not everyone should conduct research in the same way, and basic research is just one piece of that portfolio.

3

Overview of National Center for Science and Engineering Statistics Data Collections, Data Products, and Its Previous Work in This Area

Gary Anderson (National Center for Science and Engineering Statistics [NCSES]) opened his presentation by describing the work that NCSES has done in the past decade, particularly the work that relates to definitions. In that time, several definitional issues arose as academic institutions and disciplines have developed their own definitions to characterize research that reaches across disciplines. Anderson noted the transition from the use of the terms "disciplinary," "multidisciplinary," and "interdisciplinary," to the term "convergence." Each term created definitional issues and brought attention to new aspects of research that drew on expertise from various disciplines. NCSES has worked through these definitional issues in various forums including both survey development and workshops.

To describe his evolution, Anderson said that the terms unidisciplinarity, multidisciplinarity, and transdisciplinarity can reflect different degrees of integration, transdisciplinarity having the highest level of integration. The levels of integration include how the results produced relate to their individual components, how individual researchers work together, and how the process can create a more valuable product. Based on definitions in the literature, Anderson and colleagues use the following definition of convergence: an approach to problem solving that cuts across disciplinary boundaries. Two major areas are most salient for defining convergence: the deep integration of disciplines and problem-oriented research.

Deep integration can be described as researchers becoming a whole unit rather than working alone or side by side. Problem orientation can be described as an emphasis on an endpoint problem that needs to be solved, be it societal or scientific. Anderson expressed a desire to further explore these issues as convergence research continues and noted the potential it holds for broader societal impacts, as opposed to (uni)disciplinary advances.

Anderson and colleagues shared the findings of their initial background research with NCSES leadership and posed two broad questions: (1) What have NCSES leaders done, and what are they currently doing? (2) Is there a level of rigor and standardization that can be used nationally? These questions informed development of this workshop's agenda and were further divided into five more granular questions:

1. What has NCSES done that contributes to the measurement of convergence and its impact?
2. Are there NCSES programs, experiments, or pilots related to measuring convergence work that might be used as a basis for workshop sessions?
3. Is the current working definition adequate for the creation of national indicators using current surveys and reports?
4. Are current methodological approaches for NCSES surveys and indicators appropriate for the measurement of convergence?
5. Are indicators based on nonsurvey techniques (bibliometrics and scientometrics of citations, as well as network analysis and modeling) adequate for assessing convergence research and its impact, and can or should the NCSES seek to develop them?

To investigate what NCSES has done and is doing, Anderson began with decades of emails, decision documents, white papers, conference reports, and formal publications on science and engineering indicators. Regarding where NCSES is going, Anderson believes that now is a good time to reinvestigate convergence and its definition. As research becomes increasingly problem-oriented rather than merely discipline-oriented, focuses on societal impacts, and leans toward deep integration, NCSES leadership realizes the importance of measuring the impact of interdisciplinary research and of convergence.

Anderson discussed areas where NCSES is considering incorporating measures of interdisciplinarity through taxonomies and classifications of research. A 2012 workshop to discuss this possibility considered the need for NCSES to measure interdisciplinarity in the context of what the current empirical literature says about this, focusing on bibliometric datasets. The subjects discussed included methods for producing valid measurements and whether the characterizations were exhaustive and mutually exclusive, which is in turn related to apportionment issues involved with interdisciplinary research. Anderson then made a distinction between a top-down approach, which relies on existing taxonomies to measure disciplinarity (e.g., whether the authors represent various disciplines), versus a bottom-up approach that is driven by data science and machine-learning methods such as topic modeling to identify emerging fields. He mentioned the possibility of a hybrid approach that would leverage and then combine topic modeling with the priorities of the administration or Congress to produce valid measures that investigate interdisciplinarity. By the end of that workshop, attendees concluded that interdisciplinary research and development (R&D) is difficult to detect, describe, and measure, and that the existing taxonomy of disciplines does not easily accommodate an exploration of emerging outcomes. Despite these difficulties, Anderson reiterated the contemporary importance of grappling with these issues.

NCSES has been deeply involved with developing guidance for international research and design standards for interdisciplinary work. It contributed extensively to the 2015 Organisation for Economic Co-operation and Development (OECD) revision of the Frascati Manual,[1] which provides guidance on the collection and reporting of R&D performance and funding data. The manual describes fields of research and development, providing guidance on the use of taxonomies to determine levels of interdisciplinarity in a project. The manual also offers a caveat that academia may find this guidance easier to work with than other sectors of the economy. Although interdisciplinary, the experimental development work of businesses, for example, might be difficult to measure.

Anderson explained that over the course of several years, NCSES has updated the taxonomies that are used to measure research to better reflect the types of R&D conducted at educational institutions. NCSES revised its survey taxonomy to increase standardization across statistical agencies and to ensure that research fields remain up to date. Maintaining a current list of taxonomies would allow users to find their respective field more easily within a long list of disciplines presented in a survey. Other changes included adding new disciplines within examples, reclassifying some disciplines into different fields, and adding four distinctly new fields. Despite these revisions, the classification of interdisciplinary R&D has remained generally unchanged and is referred to as Other Sciences or Other Nonscience and Engineering Fields. Other Sciences is used for R&D that involves at least one science and engineering (S&E) field and also involves research expenditures that cannot be attributed to specific fields. Other Non-S&E Fields is used to categorize research and involves multiple Non-S&E Fields when a specific field cannot be reported.

Anderson also presented results from the Higher Education Research and Development (HERD) survey and explained how the survey's 2009

[1]Organisation for Economic Co-operation and Development. (2015). *Frascati Manual 2015: Guidelines for Collecting and Reporting Data on Research and Experimental Development.* Paris: OECD Publishing. doi:10.1787/9789264239012-en.

redesign investigated the possible inclusion of an interdisciplinary measure. This effort began with conversations with individual data users and universities that revealed the importance of interdisciplinarity. This led NCSES to develop a pilot survey, debrief with respondents, and create a final version, which was shown to workshop participants. To enable measurement of interdisciplinarity, the final version of the survey provided the following definition of interdisciplinary R&D:

> Interdisciplinary R&D integrates information, data, techniques, tools, perspectives, concepts, and/or theories from two or more disciplines or bodies of specialized knowledge. The purpose of interdisciplinary R&D is to advance fundamental understanding or to solve problems whose solutions are beyond the scope of a single discipline or area of R&D. Interdisciplinary research includes R&D expenditures within a center that primarily conducts interdisciplinary R&D at your institutions. It may also include R&D jointly conducted by two or more departments at your institution.

Unfortunately, the operationalization of interdisciplinary R&D did not produce a consistent measure. The inconsistencies manifested themselves in two major ways. First, different respondents within a university understood the concept of interdisciplinarity differently. Second, record-keeping methods by universities did not match the survey's definition of the concept. Rather than abandoning the effort to measure interdisciplinarity, the HERD survey instead differentiated the measurements by discipline or by broader categories. Consequently, although there is a measure of interdisciplinarity, Anderson explained that the levels of interdisciplinary research cannot be fully characterized because of differences in how interdisciplinarity is measured across disciplines.

Anderson then used the Business and Enterprise Research and Development (BERD) Survey to explain successes NCSES had with analyzing data on topics that fit its definitions of either interdisciplinary or convergence research. The BERD Survey, previously called the Business R&D Innovation Survey (BRDIS), maintains the use of definitions, but ones that are specific to a field such as biotechnology or nanotechnology. By following the guidance provided in the Frascati Manual, the survey allows NCSES to produce internationally comparable data to estimate the number of companies involved with interdisciplinary or convergent research.

The BERD survey successfully measured convergence within biotechnology and nanotechnology, but not for artificial intelligence (AI). The initial iteration of the survey raised concerns about the quality of the AI data, which could stem from the definitions provided, imperfect record keeping by respondents, or another factor. NCSES is working in collaboration with the Census Bureau to identify the source of these data quality issues.

Anderson also described the findings from the NCSES Survey of Earned Doctorates (SED). One survey question directly asked respondents whether their dissertations were based on interdisciplinary research. If yes, the survey asked respondents to identify the primary field and up to three additional fields involved in their research. Using the Frascati Manual and a 20-year time series of earned doctorates survey data, Anderson is building a measure of interdisciplinarity based on existing, field-based taxonomies. Since 2001, 24 to 45 percent of respondents indicated that at least two fields are included in their dissertation research.

In addition to taxonomies and survey revisions, Anderson investigated bibliometric data and the development of indicators for interdisciplinarity. Attempts to develop indicators in 2010 were based solely on bibliometrics and proved unsatisfactory for management and policy purposes.[2] The effort yielded inconsistent results and was thus deemed premature.

Anderson described his conclusions on the updated ability of NCSES to produce indicators and measures of inputs, outputs, and processes involved in convergence research. He began with a caveat that researchers will likely find that respondents have definitions of and experiences with convergence that differ from their own. He emphasized that the transition from articulating a good definition of convergence to operationalizing a statistical definition of convergence is thus difficult.

Anderson stated that certain research topics of convergence can be successfully measured, as evidenced by a decade of research, and that the impacts of interdisciplinary research may hold greater importance for society rather than the scientific community. The most successful example is the BERD survey on biotechnology and nanotechnology, arguably two particular topics that meet the definition of convergent research. Additionally, as was recommended in an earlier NCSES-sponsored workshop, it may be possible to implement hybrid approaches that combine top-down definitions with bottom-up approaches such as topic modeling. This hybrid approach may be able to successfully identify convergent activities within certain priority areas or topics of interest. As a highlight, Anderson referenced a partnership with researchers at the University of Virginia, who leveraged topic modeling strategies to investigate pandemics, with a particular focus on issues related to COVID-19. By extracting recent Federal RePORTER data prior to 2020, the group has been able to detect emerging trends with respect to pandemic and coronavirus research and to track topics using a convergent approach. Anderson also found that building up measures of convergence from existing taxonomies may be the most successful in the context of survey research.

---

[2]National Science Board. (2010). *Science & Engineering Indicators 2010*. Arlington, VA: National Science Foundation.

In summary, NCSES considers efforts to measure convergence research activities to be an important topic, and successes have been achieved through establishment surveys such as BERD and by leveraging existing taxonomies as has been done in current NCSES human resources statistics surveys. Further, hybrid (top-down/bottom-up) approaches may offer a potential path forward to the development of novel measurement approaches.

DISCUSSION

Nora Cate Schaeffer (University of Wisconsin–Madison) asked Anderson why data quality concerns were only attributed to AI and not nanotechnology and biotechnology. Based on administrative data and additional sources, Anderson's research team suspected that some respondents who self-reported that they were working in AI were not engaged in work that the team considered to be AI. John Jankowski (NCSES) added that nanotechnology and biotechnology also posed data quality issues initially, but the definitions were improved over time, and successive data collections were more consistent. Jankowski is hopeful that later iterations, including AI, will produce higher-quality data.

Entwisle and Ben Jones (Northwestern University) asked whether NCSES is tracking information about diversity in the R&D enterprise. Anderson responded that NCSES does try to measure diversity in the S&E enterprise, however, it remains unclear whether existing measures of diversity can capture either interdisciplinarity or convergence. Anderson also clarified that NCSES conducts surveys of enterprises (universities, government agencies, businesses) rather than individual research projects. Emilda Rivers (NCSES) commented that tracking at the project level could involve qualitative as well as quantitative data collection, which can be difficult to generalize for a national audience. She noted that measuring the impact of convergence on diversity and on ways of learning in educational institutions can speak to the Foundations for Evidence-Based Policymaking Act of 2018 (Evidence Act for short),[3] which requires the formulation of evidence-based policy. Rivers noted that NCSES explores these impacts of convergence primarily on a regional level and offers national-level information only to help a user navigate the regional data.

Entwisle clarified that as a vice chancellor for research, she had to aggregate project-level data in order to respond to a HERD survey. Jankowski confirmed that the BERD and HERD surveys aggregate project-level data that are further aggregated at the national level. He noted that additional questions regarding demographic information have been added

[3]U.S. Congress, House of Representatives. (2018). Foundations for Evidence-Based Policymaking Act of 2018 to accompany H.R. 4174. 115th Cong., 1st sess., pp. 115–411.

to the BERD survey, but although the survey asks questions about sociodemographic characteristics (sex, educational level), many respondents do not provide the level of data many investigators desire.

Joshua Schnell (Clarivate) asked whether NCSES has explored the problem-oriented aspect of convergence through bibliometric data analysis. Anderson referred this question to Beethika Khan (NCSES), who stated that no such investigation had been conducted. That investigation would require a deeper understanding of the problem or research question than such an analysis would address. She explained, though, that NCSES is attempting to link journal-level bibliometric data with data from SED to facilitate deeper demographic analysis of publications. The data currently provided by journal administrators do not allow for such demographic analysis. In responding to a follow-up question, Khan said that the data-linking process is ongoing, and so not all data are fully available to the National Science Foundation (NSF). She offered to connect interested researchers with her statistician colleagues for more information. Khan added that bibliometric data allow her team to see the ways in which countries specialize in a discipline or a topic and the kinds of collaborations that are taking place, including international collaborations and their locations. The levels of collaboration between academic institutions and businesses can also be measured.

Jankowski commented on his hope to move forward from this workshop with additional experimentation with topic modeling through natural language processing of grant documentation (in part through a partnership with the University of Virginia). He suggested that these analytic methods may reveal more about convergence than a survey instrument would be likely to do.

Khargonekar asked whether topic modeling is capable of distinguishing between problems or whether all problem-oriented research is combined, regardless of topic, in NSF's analyses. Jankowski answered that NCSES began by generating relevant search terms clustered into general topics that fall under the rubric of a selected problem. The next step is to automate this process to allow the machine to identify subtopics. Anderson added that the research team is still learning how to automate the labeling of the orientation of clusters toward problems, techniques, and other focuses.

Rivers closed the session by expressing her gratitude to Anderson for his presentation's depth and breadth and thanked all participants for their robust participation and engaging questions during the discussion.

4

Conceptualizing Convergence

As an introduction to his talk, Dan Stokols (University of California, Irvine) highlighted four consensus reports produced by the National Academies of Sciences, Engineering, and Medicine that demonstrate interest throughout the field in understanding and promoting convergence, transdisciplinary team science, and interdisciplinary research. He noted that convergence and its definitions have continued to evolve. Its hallmarks include deep integration across disciplines as well as between academic disciplines and perspectives outside academia, responsiveness to specific and compelling problems as research drivers, and the coproduction of knowledge by researchers and diverse stakeholder groups, many of which are situated outside academia. The problems that drive interest in convergence could be considered "wicked problems," a term coined by Rittell and Webber[1] to refer to interlocking problems with multiple causes for which there are no clearly established entities responsible for finding solutions. Some 21st-century global challenges that meet this definition and have been driving interest in convergence include the environment, public health, technology, and sociocultural challenges. Each of these examples could also be considered a "super wicked problem," a term coined by Levin and colleagues[2] in 2012 to describe wicked problems that require time-urgent solutions (for instance, climate change and COVID-19).

[1]Rittel, H.W., and M.M. Webber. (1973). Dilemmas in a general theory of planning. *Policy Sciences* 4(2):155–169.

[2]Levin, K., B. Cashore, S. Bernstein, et al. (2012). Overcoming the tragedy of super wicked problems: Constraining our future selves to ameliorate global climate change. *Policy Sciences* 45:123–152. doi:10.1007/s11077-012-9151-0.

A comparison between two National Academies reports, the 2014 consensus report on convergence[3] and the 2019 workshop proceedings,[4] reveals that conceptions of convergence are changing. During those 5 years, the scope of cross-disciplinary and transepistemic work broadened to encompass not only STEM fields but also the arts, humanities, and interest groups beyond academia. Researchers began to emphasize integration of the concerns and tools of transdisciplinary collaboration and convergence. They also began to explicitly consider strategies to create a culture of convergence, as well as an interplay between divergent and convergent thinking in research organizations, agencies, and college campuses. Finally, Stokols noted emerging questions about the impact of macrosocietal forces (anti-science views, the COVID-19 pandemic) on convergence, questions such as how convergence can continue as COVID-19 restrictions hinder normal opportunities for collaboration.

Stokols presented a convergence ecosystem model that situates key facets of convergence within concentric zones. At the center are individual team members, who are situated in an immediate sociospatial physical environment, such as a lab. These levels are embedded within a multiteam system such as a university campus. Each level is then affected by macro-scale influences such as national research policies and stakeholder groups. Convergence can be supported at each ecosystem level. At the levels of the individual and a team's immediate environment, it is critical to cultivate collaborative and integrative competencies. Team-specific supports may include technological and environmental resources. System-level support may include institutional incentives for cross-disciplinary partnerships such as seed grants, credit sharing, and criteria for promotion and tenure that acknowledge the importance of collaborative research. At the macro level, societal events, national research policies, and funding from agencies and foundations can positively or negatively impact opportunities for convergence. Stokols noted that social and cultural issues such as public opinion about the value of science can also affect convergence opportunities but have received less attention in the literature. (The individual, team, and institutional levels are major foci of the 2015 report *Enhancing the Effectiveness of Team Science*.[5])

---

[3]National Research Council. (2014). *Convergence: Facilitating Transdisciplinary Integration of Life Sciences, Physical Sciences, Engineering, and Beyond.* Washington, DC: The National Academies Press. doi:10.17226.18722.

[4]National Academies of Sciences, Engineering, and Medicine. (2019). *Fostering the Culture of Convergence in Research: Proceedings of a Workshop.* Washington, DC: The National Academies Press. doi:10.17226/25271.

[5]National Research Council. (2015). *Enhancing the Effectiveness of Team Science.* Washington, DC: The National Academies Press. doi:10.17226/19007.

Stokols explained that the 2015 report defines the field of team science as an interdisciplinary field concerned with understanding and managing the circumstances that facilitate or hinder the effectiveness of cross-disciplinary research, training, and translational initiatives. The literature on team science has grown significantly since 2006 and provides a vast toolbox of practical strategies for improving the effectiveness of collaborative research. When considering measurement, team science offers strategies for evaluating both interdisciplinary and transdisciplinary contributions. When using bibliometric measures to gauge team productivity and impact, supplementary qualitative data are generally needed to understand the catalytic impact of the ideas generated by research teams. A team may produce many bibliometric outputs for example, but the incremental significance of each publication may be low compared with the significance of a ground-breaking publication. Interdisciplinary and transdisciplinary scholarly contributions may represent somewhat different categories of convergence in that transdisciplinary collaborations by definition entail deeper integration across multiple fields and more novel or catalytic breakthroughs. A wide range of nuance and measures needs to be considered when measuring different forms of convergent research. The field of collaborative transdisciplinary scholarship and practice provides an array of tools and metrics to measure the different facets of the convergence ecosystem.

Stokols listed several practice resources for team science such as the National Institutes of Health (NIH) National Cancer Institute (NCI) Team Science Toolkit. The University of California, Irvine (UCI) considers "team scholarship" to include fields outside of STEM and provides resources through its Team Scholarship Acceleration Lab. The NIH NCI Field Guide led by Michelle Bennet and her colleagues is a useful hands-on toolkit that provides strategies for effective collaboration in team research settings. Michigan State University developed the Toolbox Dialogue Initiative, which aims to bridge the philosophical differences that may arise among research team members. Finally, the Swiss Academies of Arts and Sciences Toolbox addresses an array of strategies to advance and measure the coproduction of knowledge in and beyond academia.

Stokols described two important facets of cross-disciplinary collaboration and convergence: (1) fostering intellectual synergy and diversity of thought, and (2) promoting a social integration of research teams and centers resulting in mutual coordination over several phases. The field must consider how to promote each of these facets in the context of the macrosocietal forces that affect the problems that researchers aim to solve. Prior to the COVID-19 pandemic, for example, practitioners understood proximity and environmental supports to be necessary for interpersonal collaboration. Stokols posed a question: How will the reduction of co-location during the COVID-19 era affect the convergent capacity of research teams and centers, and what new competencies of convergent research will become imperative for carrying out research? Attention should furthermore be paid to the potential effects on training in convergence that may have shifted during the pandemic.

Stokols continued by providing examples of environmental supports for convergence. He encouraged brainstorming exercises that incentivize divergent thinking, which can be adapted for online use. Although some aspects of convergent research can accommodate remote settings, Jason Owen-Smith and colleagues find that co-location in the same building and overlapping activity zones (coffee rooms, break areas) contribute to the likelihood of effective collaboration and of winning cross-disciplinary grants. Research teams and centers often highlight and share their collaborative identities in shared spaces by posting photographs of team members, conference posters, and certificates of collaborative awards. Using shared physical spaces to communicate collective values reinforces collaboration and convergent behaviors. The reduced availability of shared spaces and in-person meetings must be considered as team scientists adapt to the pandemic, during which remote collaboration has become increasingly common.

Stokols outlined four core competencies that convergence training seeks to produce. Convergence must be transdisciplinary, team-based, translational, and transcultural. A normative competence instilled in people who are engaged in the coproduction of knowledge and translational work guides allocation of resources and effort. The coproduction of knowledge may be particularly complex in international collaborations, as the additional political and cultural differences may inhibit convergence.

To close the presentation, Stokols asked attendees to consider the first four National Academies' consensus reports on convergence: *Facilitating Interdisciplinary Research*[6] (2005); *Facilitating Transdisciplinary Integration of Life Sciences, Physical Sciences, Engineering, and Beyond* (2014); *Enhancing the Effectiveness of Team Science* (2015); and *Fostering the Culture of Convergence in Research* (2019). He advised the attendees to consider what questions they would like to see explored in the next report: Will there be standard or context-dependent definitions and criteria for convergence. Will a typology of convergence emerge where some forms may include a coproduction of knowledge among academic and nonacademic partners? How will competencies for convergence shift in the COVID-19 era? Finally, can we transcend spatial, sociocultural, and geopolitical constraints on convergence?

[6]National Academy of Sciences, National Academy of Engineering, and Institute of Medicine. (2005). *Facilitating Interdisciplinary Research*. Washington, DC: The National Academies Press. doi:10.17226/11153.

DISCUSSION

Klein asked how Stokols' work informs the tension between openness and closure, and what implications might follow from the Steering Committee's definition of convergence. Stokols suggested that if convergence is viewed as an umbrella concept that encompasses interdisciplinarity, transdisciplinarity, community engagements, and the coproduction of knowledge, then the creation of a global, standardized measure of convergence of either the quantitative or qualitative variety will be challenging. A taxonomy of convergence may be more realistic, and certain facets of convergence may be emphasized within different research initiatives or collaborations. Convergence involves several interrelated dimensions, and Stokols stressed the difficulty inherent in imposing a single standard definition. Within a single research team, the measure of outputs will change based on the research phase. In the initial phase, researchers focus on creating new ideas and collaborative processes, so bibliometric and other outputs are few. In later years of the collaboration, researchers begin to generate products, but the number and timing of those vary according to different types of teams. Scholarly teams, for example, may seek new knowledge, the value of which depends on qualitative analysis and appraisal. This type of output differs substantially from that of a manufacturing or corporate team. It is often more difficult and takes more time to gauge the novelty and value of intellectual outputs created by scholarly teams than it does to assess quantity or utility of products generated by nonacademic teams.

Holbrook asked whether a plurality of definitions complicates the prospects for measurement. Stokols responded that measuring a complex phenomenon requires confrontation and identification of its dimensions. In his view, the various facets of convergence at multiple levels must be measured, an approach that is facilitated by the ecosystem model.

Erin Leahey (University of Arizona) asked about the relationship between diversity and the difference between divergent and convergent thinking. Stokols believes the two topics are closely related. To be productive, effective, and creative, a research team must first leverage the diversity of thinking within itself. Research teams and centers may become preoccupied with obtaining and renewing funding and thus may fail to allot adequate time for brainstorming and divergent thinking. Stokols emphasized that this kind of work should exist in a running dialectical exchange with other components of convergent thought such as efforts to integrate fields by bringing research groups together to emphasize common ground. Referring to Klein's question about navigating the relationship between diversity and the differences between divergent and convergent thinking, Stokols suggested that the variety of categories of diversity (skills, knowledge, and culture) enhance diversity of thinking. The correspondence between the diversity of a team's composition and the divergence of the ideas it produces may not be one-to-one, but they are positively correlated—provided that the team's diversity is managed effectively to avoid fissures (or "faultlines") that can develop between team members from different backgrounds.

Khargonekar asked whether lessons learned from the measurement of ecosystems might inform the Steering Committee's deliberations about how to move convergence research forward. Stokols suggested that systematic thought across multiple scales is one core competency for convergent work. Many methodologies are available for ecosystem measurement, but a fuller discussion, he said, falls outside the scope of this convergence workshop.

Entwisle asked Stokols to imagine a potential typology of convergence. Stokols explained that such a typology would have many organizing dimensions such as the extent to which academic fields and nonacademic cultures are involved in the coproduction of knowledge; the question of whether the desired outcome relates to policy making, a medical product, or scientific knowledge; and the potential for near- or long-term applicability. Some convergent work may have results that can immediately be translated into policy, whereas other research may be intellectually novel and catalytic but not have immediate policy implications.

5

Convergence as Practiced

The panel discussion summarized in this section brought together experts who have studied and overseen collaborations in a variety of settings and can provide their perspectives on what the National Center for Science and Engineering Statistics (NCSES) may usefully measure. Lora Weiss (Pennsylvania State University), the first panelist, discussed some implications of convergence. To recognize and remind others about the importance of continuing work on convergence technologies, she began by quoting the National Strategy for Critical and Emerging Technologies (C&ET) October 2020 report, which stated that "[i]n fact, many technology breakthroughs occur at the intersection of two or more disparate technologies,"[1] The National Strategy for C&ET emphasized interdisciplinary work and the need for academic, public, and private partnerships to address societal challenges.

The Journal of Higher Education recently published an article[2] questioning whether hiring initiatives at U.S. research institutions that focus on interdisciplinary clusters represent an ideal rather than delivering actual collaborative relationships. Commenting on the authors' highlighting of Penn State as an exemplar institution, Weiss explained that the interdisciplinary research institutes at Penn State were established based on existing and genuine partnerships between interdisciplinary institute leaders and deans

[1]The White House. (2020). *National Strategy for Critical and Emerging Technologies*. Available: https://www.whitehouse.gov/wp-content/uploads/2020/10/National-Strategy-for-CET.pdf.

[2]Bloom, Q., M. Curran, and S. Brint. (2020). Interdisciplinary cluster hiring initiative in U.S. research universities: More straw than bricks? *The Journal of Higher Education* 91(5):755–780. doi:10.1080/00221546.2019.1688615.

of academic colleges, all of whom shared expectations and a vision for the center. This shared outlook, Weiss noted, is crucial: If researchers remain within a single discipline, interdisciplinary work cannot function.

The institute directors are furthermore top scholars in their area of expertise, which helps in the recruitment of junior hires. Incoming researchers are more likely to be interested in interdisciplinary work when they know that the director has experienced the tenure process and understands the experiences of junior faculty and that such work will not harm their promotion paths.

Each of Penn State's core research facilities is staffed by research professionals, which constitutes an important component of the institution's success. These dedicated staff allow researchers to focus on their research rather than become absorbed by activities such as maintaining and operating equipment. The final element of Penn State's success is its seed grants. The university invests in high-risk, high-payoff collaborations that lead research in new directions and help to foster new partnerships. Roughly half of the projects receiving Penn State seed grants receive external funding. Weiss cautioned that if this ratio were to approach 100 percent, it would signal that the supported projects are not taking enough risks.

To demonstrate the range of interdisciplinary collaboration at Penn State, Weiss presented a diagram that shows all internal Penn State collaborations (spanning 24 campuses and other extensions) that have generated at least 15 shared publications. The university's external partnerships further extend its collaborations and allow diverse perspectives to flourish through interactions and relationships that Weiss and colleagues constantly fine-tune.

In his talk, Peter Schiffer (Yale University) spoke about administrative structures and incentives that encourage convergence. He noted that measures of convergence outcomes are difficult because they try to measure ideas and impacts that are brand new, requiring researchers to work to measure things almost before they exist. On the other hand, he observed that convergence can be fostered when an institution's culture allows it, which depends on administrative structures, policies, and incentives. Therefore, structures and incentives within an institution may be surveyed to provide indirect insight into whether an institution is creating an environment to promote and support convergence.

Schiffer described several specific administrative structures that create such an environment. First, support for diversity, equity, and inclusion in research is critical because diverse teams produce better outcomes, particularly in areas that do not represent a single, well-defined discipline. Large institutions and centers that reside outside of disciplinary centers and allow people to collaborate across disciplines are also important. Collaborations can be promoted even further through campuswide events, talks, or research sessions that informally bring together researchers from different fields and facilitate social and professional interaction. Schiffer reiterated Weiss's observation that professionally staffed, centrally managed core facilities make it easier for people to meet and collaborate using the same instruments to develop new ideas.

Funding mechanisms are also important. A critical mechanism is an accounting system that effectively manages funds across multiple units. If a single person in one department controls all funds, interdisciplinary teams may find it difficult to work collaboratively. Universities can similarly help build a fundraising arm for transdisciplinary work that can help start or seed new projects. Traditional university fundraising is ill-equipped for this work because it is organized around alumni. Transdisciplinary convergent efforts typically do not have alumni because they do not grant degrees and are relatively new. Movement toward convergence is facilitated by offering interdisciplinary degrees to students. Degree programs can bring students together across disciplines beyond the obvious connections between biology and chemistry or between physics and math. Students can also create bridges that join faculty across disciplines.

Beyond structure, Schiffer outlined policies that support convergent research. Tenure and promotion policies are a particularly difficult case because most universities depend on external reviewers selected for their disciplinary expertise. Convergent work can be recognized by explicit policies in the tenure and promotion process. Graduate students should be able to work with different or multiple advisors, regardless of their admission department. Postdoctoral students and research scientists should likewise be able to hold appointments that cross units. Internal funding policies can also support convergent research by ensuring that individuals are not forced to compete with each other for things like research incentive funds associated with indirect cost reimbursements, but instead are rewarded when convergent activities take place. Finally, Schiffer addressed incentives, particularly for faculty members. If internal, generous, and stable funds support convergent research, then faculty will pursue convergent research opportunities. Offering faculty discretionary funds to investigate their newest ideas will provide creative researchers with the opportunity to try new things in a manner not directly dictated by a dean or department chair. Also important is dedicated staff to assist in proposal writing and grant management. Public relations activities that celebrate the productive crossing of disciplinary boundaries can also incentivize researchers, particularly when participating researchers are recognized instead of just specific departments.

Schiffer ended his talk by adding some caveats to his suggestions. First, research on the ideas presented continues to evolve and should be studied over time. Even successful ideas need a supportive institutional culture to flourish. If the leadership culture does not support convergent research, it will not happen regardless of incentives, structures, and policies. He acknowledged that many of his suggestions are aimed at larger institutions and single institutions. Tailored approaches may be required for smaller schools and an interinstitutional convergence measure may be needed to capture activities occurring across institutions. Finally, Schiffer cautioned that not all research should be convergent. Convergence depends on disciplinary advances, and incentives and policies that promote convergence should not disadvantage researchers who work best within their specific discipline and whose work may ultimately lead to a disciplinary breakthrough or a convergent effort.

Dan Gallahan (National Cancer Institute) opened by stating this workshop has underscored the importance of understanding how we do science. As a funder of science in the Division of Cancer Biology at the National Cancer Institute (NCI), he has seen billions of dollars toward basic cancer research move through his office. At various points in time, he has used all the different terms spoken during the workshop: team science, integrated science, interdisciplinary, transdisciplinary, and others. Unfortunately, many of the research efforts launched before these terms were fully defined, and though they are related, they do differ in meaning.

Based on his background in cancer research, he noted that several aspects of convergence in science relate to cancer and our scientific understanding of it. Gallahan said that cancer incidence and mortality rates have remained steady. About 20 years ago, NCI realized many of its efforts did not hold or address the same view of cancer as other research communities. New approaches to cancer research were needed, which incentivized an incorporation of the ideas of convergence, team sciences, and integrative biology into current research. These approaches increase the variety of perspectives and abilities available for use in investigating imaging and the different scales involved in cancer research.

Gallahan stated that NCI has established two programs to broaden perspectives on cancer: the Physical Science Oncology Network and the Cancer Systems Biology Consortium. Gallahan explained that multiple viewpoints can come together through a combination of enhanced insight into a topic and efficient problem-solving skills. To begin this work, NCI convened meetings of investigators, physicists, engineers, and mathematicians to share their perspectives about the complex problems that exist in cancer research and ways which these problems can be addressed. The aim of the Physical Science Oncology Network is to identify new ways to view these problems rather than to solve these problems outright. The Cancer Systems Biology Consortium brings together a different set of communities that focus on computational approaches to enhance understanding of cancer's development and to model the body's immune response to cancerous cells.

This goal requires input and expertise in modeling and computational biology as well as in technology.

Both programs were very successful beyond the standard bibliometric measures of output. Each project involved scientists from multiple disciplines and required (based on proposal requirements) an educational and outreach component. Investigators were retrained to ensure that they possessed the vocabulary necessary to communicate across disciplines. Each program also ensured that a new generation of researchers entered the field.

Gallahan emphasized a need for critical evaluation of the programs to determine their success and future directions through engagement of an external expert panel and extensive quantitative and qualitative data collection. Evaluation should consider the diversity of publications, the extent to which the principal investigators represent multiple disciplines, the use of new terminology, and novel grant applications (incorporating new investigators with applications in new areas). Another measure of success could be the development of new programs such as the spread of computational modeling at the National Institutes of Health (NIH), which was almost nonexistent at NIH only 10 years ago. Gallahan is encouraged by the training that has been developed for young investigators and the high quality of cross-disciplinary collaboration by those new investigators.

Maryann Feldman (University of North Carolina at Chapel Hill) was the final panelist of the session and used her time to represent the research done by a group of scholars who were formerly known as the Science of Science and Innovation Policy (SciSIP) Program, and now reorganized as the Science of Science Discovery, Communication, and Impact (SoS:DCI). The group seeks to understand how knowledge is created and transformed into socially and economically useful products. Feldman herself is most interested in this work's spatial dimension, an interest that springs from her background in economic geography. Economic geography itself is a convergent topic that investigates the spatial organization of economic activity as well as complex relationships and emerging activity patterns within a place.

A great deal of work is being done on the emergence of new academic disciplines, varieties in organizations, and the creation of entirely new industries. The idea of an ecosystem is key to this understanding, and it involves the set of institutions and constituent factors that enables creativity and legitimacy. The question is the emergence of something distinct and new, even as it blends other existing types of expertise.

Feldman stated that the emergence of new academic disciplines and departments is a clear signal of legitimacy. An academic discipline requires accreditation, which is a formalized professional process. Feldman's group investigated the emergence of biomedical engineering as a distinct discipline, and preliminary findings suggest that a university requires a strong medical school as well as an engineering school, combining these strengths with the ability to bring in large research grants. Unfortunately, if the university is a Historically Black College or University (HBCU), then the probability of a biomedical engineering department's emergence decreases. The university's geographic proximity to firms that use biomedical knowledge increases the likelihood of department creation. Specifically, the density of research firms within a 60-mile radius is directly related to the probability of a university adding a new biomedical department.

Emerging industries can also be tracked using the technology categorization scheme used by the U.S. Patent and Trademark Office (USPTO). The USPTO categorizes patents and plans to establish a new patent class for emerging fields. The categorization allows researchers to see what firms and universities are developing inventions in a particular area and location.

She closed her comments by expressing the state of convergence and the opportunities that it presents. In the COVID-19 era, a time characterized by a broader recognition of societal problems, business schools are pondering the timing to redirect strategies away from shareholder gains toward stakeholder and user needs. Feldman highlighted the importance of underlying values, including sustainability and social justice.

DISCUSSION

Klein asked whether each presenter could connect their presentations and findings to the previous presentations, particularly those of Dan Stokols (University of California, Irvine) and Gary Anderson (NCSES). Holbrook added a request that the panelists address how their own experience practicing convergence might inform a definition that could be used by NCSES.

Weiss said that it would be ideal to have consistent language for distinguishing between the terms "interdisciplinary" and "convergence." Consistency across universities will be particularly critical as the tenure process considers convergence. Penn State uses the term "interdisciplinary" throughout the institution. All interdisciplinary institutes comprise faculty from various departments, and hiring is done collaboratively between the institutes and university deans. The practice of cohiring allows departments to gain faculty members for disciplinary advances, while new hires allow the interdisciplinary institute to broaden its base of expertise.

Gallahan agreed that a clear definition is useful when evaluating and understanding metrics, but he does not feel hindered by the lack of a clear definition at present. As faculty are encouraged to engage in team science and interdisciplinary work, Gallahan said challenges might arise as junior scientists and postdoctoral students advance through their careers. These scientists may be part of different programs but may feel compelled to follow a more classic trajectory of publishing first-author works in a narrow way to line themselves up for promotion. Gallahan said that funders need to recognize that they are part of the problem. A researcher may put forth impactful interdisciplinary work that nonetheless may be difficult to justify to a traditional tenure promotion committee.

Schiffer added that the definition depends on the goal. Why does NCSES want a definition of convergence? Will the measure be used to determine whether convergence is happening or to measure the impact of convergence research? Schiffer stated that slight variations of the definition may be needed, from which different measurements would follow.

Feldman noted that NCSES provides data measuring university research and development (R&D) organized by traditional discipline, but it does not yet consider research center activities and the contributions made within universities.

Holbrook referred to earlier presentations that noted a desire to measure convergence because it may have a greater societal impact, while disciplinary research may have a more significant scholarly impact. The different practices and outcomes of convergence may bring about solutions to society's most pressing problems, which constitutes a motivation for measuring convergence. The flourishing of disciplinary scholarship should continue, but convergence must be the focus because it promises greater societal impact, which will influence its definition, Holbrook said.

Gallahan asked Holbrook to clarify whether he was referring to measuring how well convergence research solved societal problems in the past, or how much convergence research is currently occurring within an institution. Holbrook stated he did not know the answer to this exact question, but he believes that the first priority of NCSES should be the need to understand whether convergence can be measured, and if so, how to best carry out measurement.

Weiss reiterated Schiffer's previous point that the traditional, single-focus disciplines cannot and should not be forgotten or discounted, as interdisciplinary research depends on advances within these disciplines. She also maintained that the best of disciplinary fields come together, and that each investigator's contribution must be recognized. Gallahan agreed that modes of recognition could change, but he questioned who in academia is paying attention to these changes. Gallahan added that such changes may be suitable for more senior investigators who have already established their credentials, but unconventional modes of recognition might be detrimental to junior and untenured investigators. The responsibility to acknowledge interdisciplinary contributions falls to both academic institutions and funding agencies. Gallahan addressed a balance between focused research and more interdisciplinary programs. He views NCI as heavily weighting Research

Projects Grants (R01), so providing metrics for interdisciplinary programs could assist in quantifying the success of future interdisciplinary work.

Feldman spoke about typologies, referencing Pasteur's quadrant.[3] This classification distinguishes between basic and applied science, with the most fruitful topics of investigation having an applied orientation that adds new knowledge. Using academic entrepreneurs as an example, she argued that the most productive academic scientists are those engaged with commercial activity that translates their research into practice. Academia brings in creative graduate students but then trains them within a disciplinary box. Consequently, this practice could cause a potential decrease in creative productivity of recent doctoral graduates.

Gallahan concurred with Feldman and noted the increasing tension between supply and demand. At NCI, application rates have nearly doubled in the last 5 years. If the budget does not match that rate, then award rates will flatten. Additional considerations need to be made for collaborative research and bibliometric acknowledgment. Researchers are less likely to have first-author publications as their collaborative work increases, which may prevent them from advancing within academia.

Nora Cate Schaeffer (University of Wisconsin–Madison) asked about what happens as convergent teams and their membership shift over time. Do the groups tend to stay together and investigate different problems, or do individuals leave a group to join a different group based around a different problem? Weiss replied that research funded with partnership-forming seed grants continue to work together. At Penn State, the relationship is seeded along with the research. Anecdotally she finds that once an investigator has collaborated in interdisciplinary research, they become more open-minded and willing to collaborate with other researchers in the future for different projects. Schiffer and Gallahan agreed with Weiss's statement, and Schiffer added that institutional support for working together in teams is critical. Strong support for working together in interdisciplinary teams makes future research feasible.

Andrew Zukerberg (National Center for Education Statistics) asked whether areas of convergence can turn themselves into new disciplines. Feldman responded that this is possible, pointing to the example of neuroscience. Initial work began with the limitations of neurology and searches for new tools to investigate problems. She said that more work is needed in this area to allow new disciplines to blossom. Ideas have previously been described as "sleeping beauty science," that is, fundamental ideas lying dormant until an ecosystem concept is applied that allows a quorum of understanding to emerge. Business demands or funding calls are often the

---

[3]Stokes, D.E. (1997). *Pasteur's Quadrant: Basic Science and Technological Innovation.* Washington, DC: The Brookings Institution.

driving force to introduce a new discipline, but this method often does not consider larger social implications.

Stokols responded to Feldman's comments to consider Pasteur's quadrant and the value added when research outcomes have near-term applicability. A longer latency period between the initial concept and the buildup of empirical evidence can be translated and applied in other areas over time. Stokols suggested that investment in convergence research that investigates basic work could have translational applicability in the future. He suggested distinguishing between different types of convergence when developing a taxonomy. Gallahan noted that Pasteur's quadrant is a constant challenge across all disciplines and said that he often finds himself defending basic research.

Josh Schnell (Clarivate) asked whether a national measure's utility depends on how a research portfolio balances, for example, basic, applied, and translational research. Schiffer pushed back on the concept of a single national measure of convergence, stating that a portfolio of metrics is more likely to be of value. Feldman agreed with this sentiment.

Schaeffer asked whether convergence could include translational research. She also asked whether social sciences and the humanities are considered niche areas dependent on a specific problem. Gallahan answered that translational research could be combined with other types of research. Schiffer agreed with Gallahan regarding translational research and added that social sciences and the humanities are not niche fields. Historically, the humanities have not been thoroughly integrated with the STEM fields at universities, but humanities scholars can play necessary and important roles that enhance the value of STEM research and vice versa.

Weiss expressed skepticism of the idea of a single definition and metric for convergence. Holbrook questioned whether Weiss would be more willing to accept a single definition of convergence if it were measured in multiple ways. Weiss acknowledged that she could accept a single definition with multiple measures, but only if the definition captured nuance, which could be very difficult. Holbrook commented on the work that Anderson presented, in which web survey participants are shown a single definition but can use a hyperlink that provides more detail and multiple examples. He asked whether providing additional definitional information in the hyperlink could still be considered providing a single definition, which Weiss found acceptable.

Schiffer added his description of convergence, which provides a framework that recognizes the importance of addressing several facets of research, particularly team dynamics and institutional incentives for facilitating collaboration, in order to accomplish an overarching aspiration rather than a monolithic goal. Gallahan agreed with Schiffer's description and endorsed that definition of convergence.

6

Outcomes and Impacts of Convergence

Erin Leahey (University of Arizona) shared findings from her investigation of the outcomes and impacts of interdisciplinary research, which she refers to as IDR. She described the impacts at three different levels: the person level, the team or publication level, and the organizational level. Her own work focuses primarily on the organizational level, emphasizing not only the outcomes and impacts of IDR, but also its unique measurement on this level.

Based on interviews with fellows in the Mellon Foundation New Directions program, her earlier book chapter focused on IDR at the person level. Study participants were all scholars who had received doctoral degrees and tenure in a particular field and applied for the fellowship to pursue a different field. Leahey and McBee[1] (2016) focus specifically on the challenges faced by participants. Common themes included production and evaluation hurdles. Production hurdles described by the participants included the extra time and greater commitment that IDR requires, which reduces one's productivity. Leahey highlighted one participant's response: "If one looks at my research output ... there's actually a big gap. It looks like I went into a coma or something."

An evaluation hurdle shared by study participants was the need to placate disciplinary colleagues. Participants felt required to capture the interest of scholars from multiple disciplines and thus faced different types of criticism from each discipline. Participants perceived they were being

[1]Leahey, E.E., and D. McBee. (2016). New directions in interdisciplinary training: trials and tribulations. In *Investigating Interdisciplinary Research: Theory and Practice across Disciplines*. Rutgers, NJ: Rutgers University Press.

judged by the standards of others. Leahey said that all these participants held tenure in the humanities disciplines like philosophy and English in which sole-authored work is common. She wondered whether other disciplines in which team publication is more common would face the same challenges.

To address the team or publication level, Leahey described her 2017 article[2] in *Administrative Science Quarterly* that she coauthored with Christine Beckman and Taran Stinko. This publication investigated 30,000 papers written by more than 900 authors across a variety of fields, collected from scholars at the Industry University Collaborative Research Center (IUCRC). The measure of interdisciplinary research was based on the 2007 Web of Knowledge Science Categories devised by Alan Porter and his colleagues[3] to measure the level of integration. To measure how interdisciplinary a focal paper is, they counted the number of references cited. Leahey and her colleagues assigned a score between 0 and 1. The results suggested that interdisciplinary research imposed a productivity penalty, but it had a positive effect on visibility (as measured by number of citations). In terms of effect size (based on the standardized coefficients), the productivity penalty was greater than the boost in visibility.

According to Leahey, interdisciplinary research and collaboration pose cognitive challenges that contribute to the productivity penalty. It becomes more difficult to generate ideas, and research moves at a slower pace. Survey data also document collaborative challenges, including initial difficulty communicating within interdisciplinary teams. With repeated collaboration, though, coauthors build a comfort level that results in more productive collaboration. Leahey and her colleagues did not find that interdisciplinary research posed additional challenges during the review process. Based on working papers from archive.org, interdisciplinary papers were more likely than others to eventually be published.

Visibility for IDR research also depends on the fields represented. Life sciences, for example, which is considered a highly interdisciplinary field, experiences a more significant positive impact on author visibility. Leahey said that the effects of IDR are also greater when the cognitive distance between interdisciplinary fields is larger. Leahey made a distinction between cognitive distance and variety: when variety is the focus, the benefits of visibility are weakened, though the production penalty disappears. IDR is also risky in the sense that is increases the variability of visibility (as measured by the standard deviation of citations across all the scholar's work).

---

[2]Leahey, E., C.M. Beckman, and T.L. Stanko. (2017). Prominent but less productive: The impact of interdisciplinarity on scientists' research. *Administrative Science Quarterly* 62(1):105–139. doi:10.1177/0001839216665364.

[3]Porter, A., A. Cohen, J. Roessner, and M. Perreault. (2007). Measuring researcher interdisciplinarity. *Scientometrics* 72:117–147. doi:10.1007/s11192-007-1700-5.

With funding from National Science Foundation (NSF) and Science of Science and Innovation Policy (SciSIP), Leahey partnered with Sondra Barringer to investigate IDR at the organizational level (University Commitment to Interdisciplinary Research: Scope, Causes and Consequences). This research focuses on structural commitments to interdisciplinary research, terms of funding, the organization of units within the university, and the question of whether or not these structural commitments bolster IDR. Units might be reorganized, for example, to add a new department or program or create a research center that connects multiple fields. Leahey and her colleagues collected data on roughly 160 Research I universities and analyzed the list of departments, names of those departments, and research centers. They used semi-supervised machine learning to identify which were interdisciplinary.

Their findings highlight four distinct indicators of a university's structural commitment to IDR: (1) the ratio of research centers to departments, (2) the ratio of interdisciplinary research centers to research centers, (3) the ratio of interdisciplinary departments to all departments, and (4) the number of departments. Using traditional confirmatory factor analysis, Leahey and Barringer devised a score that represents the university's commitment to IDR. They then used that score as a predictor to model research outcomes. Their conceptual model assumes a lag between the university's structural commitment to IDR (measured in 2012–2013) and the effect on subsequent publications and grants (2015). They found that a structural commitment to IDR has a large effect on research activity, particularly for IDR. A moderate effect was found for interdisciplinary publications and interdisciplinary National Institutes of Health (NIH) grants, but they found no effect for interdisciplinary NSF grants.

Leahey speculated about the source of the discrepancy between the effects on NIH and NSF awards, noting that NSF grants are typically awarded to a single principal investigator or to multi-university teams. This may explain why the infrastructure within a particular university plays a lesser role than it does for NIH, which is more likely to award institutional grants that involve multiple departments within the same university.

They found that structural commitment to IDR had no measurable effect on research quality (number of highly cited researchers, number of articles published in *Nature* and *Science*). She concluded that universities seeking to promote high-impact research may need to move beyond structural commitments to interdisciplinary research and express their commitment to IDR in other ways.

In another study, Leahey and colleagues investigated the precursors of a university's commitment to IDR. They used a wide variety of sources including National Center for Science and Engineering Statistics (NCSES) datasets, the Survey of Earned Doctorates (SED), the Higher Education

Research and Development (HERD) survey, the Integrated Postsecondary Education Data System (IPEDS), and NSF/NIH award data. She described both top-down and bottom-up influences that determine structural commitment to IDR. To quantify top-down support for IDR, they analyzed textual data from speeches delivered by the university president. Then, they compared the measure of presidential rhetoric supporting IDR with data from the survey of earned doctorates to test whether it predicts the number of interdisciplinary doctorates the university graduated. Top-down administrative efforts and bottom-up efforts from faculty both promote a structural commitment to IDR as the number of interdisciplinary doctoral graduates suggests.

In summary, Leahey concluded that individuals and teams benefit from interdisciplinary research through increased visibility as measured by citations, although they may experience decreased productivity. At the organizational level, commitments to IDR increased interdisciplinary research activity in terms of publications and NIH grants.

Ben Jones (Northwestern University) described the importance and difficulty of convergence and the policy design principles that follow from these characteristics. Jones noted that while some of his economics colleagues believe research is suffering a dearth of ideas, he believes that ideas such as those about the human lifespan, climate change, nuclear fusion, artificial intelligence (AI), and more are instead becoming more difficult to explore. The problem may lie not in the ideas needing research, but in the tools available to the researcher.

Jones attributed this difficulty in part to the burden of knowledge. Past research has amassed a large volume of insights, facts, and conceptual theories about the universe, but a single person has a finite knowledge capacity. As individuals, our ability narrows. To invest the time to become an expert in an increasingly complex area, a researcher's focus must be narrow. Jones argues that as individual expertise narrows, teams become essential vectors to drive scientific advancement on large-scale problems. An assembly of diverse and complementary expertise is essential to achieve breakthroughs. Although teams are not synonymous with convergence, they are related, constituting aggregations of people with a variety of expertise.

Compared to sole authors, the importance of teams is illustrated by their advantage in producing the highest-impact work as measured by citations received, a phenomenon that represents a reversal from the 1950s and 1960s. Jones investigated this advantage by considering the relationship among teams, citation practices, and the probability of a groundbreaking publication. He began by using the set of references cited by a given publication as a measure of novelty. That set could be scored for how commonly its particular network of references appeared together across published works, and he labeled uncommon combinations as novel. Publications that balanced conventional and novel references had a higher probability of becoming highly cited and influential. Publications with only common combinations have a low probability of success, as do those with too many novel combinations. Additionally, teams have an increasing advantage in achieving this balance and therefore in developing high-impact ideas compared to sole-authored publications.

This dynamic also holds true relative to the "age" of citations (the length of time between the release of a cited publication and its citation in another publication). Jones found that the most influential papers balance extensive citation of the most recent literature (within the past 5 years) with variable citation of older papers (10 to 15 years old or more). Achieving this balance requires appreciating long-standing knowledge while simultaneously being grounded in a constant influx of new information. Although an individual specialist can achieve this balance within one field, that individual will likely not be able to achieve such a command of the literature across many fields. Indeed, teams again have an advantage in achieving this kind of chronological balance.

Although Jones' research suggests that teams are critical to producing knowledge, that observation raises the question of how to facilitate good combinations that allow groups to work together effectively. He suggested that research and development (R&D) outcomes are difficult to anticipate, as exemplified by the poor predictive record of venture capitalists. The challenge is confounded by the narrow expertise of individuals, reviewers, and funders of research: Each individual is constrained by his or her own blind spots and areas of expertise. Jones added that the potential for research combinations is effectively infinite, although most combinations are likely to be useless.

In closing, Jones presented tentative design principles for high-impact convergence work. Jones highlighted a focus on demand as critical and relatively easy: Researchers have an ample number of problems they would like to solve. Defining which problems to solve is a starting point, and agencies such as NSF can play a major role in guiding this selection. For basic research, a source of funding that does not require a monetary return on investment is ideal. This approach ensures that researchers can focus on knowledge production rather than product creation. Although others might choose the selection of problems, Jones suggests allowing teams to self-assemble and letting their work converge organically. Unforced collaboration is likely to support better communication and compatibility. As a final suggestion, Jones suggested that a research portfolio should expect and plan for failure by creating a portfolio of diverse avenues of problem-solving and coordinating efforts rather than pretending to control outcomes. Failure is a regular occurrence within R&D, although there will be political and economic pressure to avoid investing in endeavors that may fail.

DISCUSSION

Following the individual presentations, Holbrook asked Leahey about the extreme polarity between the "hits and misses" experienced in attempting to publish interdisciplinary research findings. Holbrook noted, for example, that when he has submitted grant applications, reviewers frequently markedly gravitate toward extremes, half of them rating the proposal as excellent and the other half rating it as poor. He wondered aloud whether Leahey sees bipolarity as a potential indicator of interdisciplinarity, convergence, or potentially transformative research. Leahey acknowledged that anecdotal evidence suggests that Holbrook's experience is not unique, but she did not have data that could address the question. She noted that her data have suggested that the peer review process for interdisciplinary publications is no more onerous than for monodisciplinary work.

Bethany Laursen (Michigan State University) added a related comment about the use of standard deviation to capture some features of bibliometrics. She suggested that quantitative researchers could use some derived measures rather than direct measures to understand aspects of diversity and convergent versus divergent thinking. Owen-Smith responded that using a portfolio framework to carry out research would allow those derived measures to be calculated over a broader distribution of collected data.

Khargonekar asked Jones whether collaborative success is based more on random luck than expert judgment. Jones suggested that expert judgment matters a great deal but is difficult to measure. He also emphasized that while some researchers may celebrate team science as inherently valuable, he sees it simply as a necessary solution to the brutal problem of the increasingly narrowed focus in various disciplines. Moreover, he noted that teams have their own problems stemming from frictions between diverse perspectives and the implicit and intuitive knowledge held by each team member that can never be fully accessed by others. The need for the team approach in academia, though, requires that these challenges be worked out organically.

Ismael Rafols (University of Leiden) suggested that evidence may not support the use of citations as a proxy for societal impact in applied fields and asked the entire panel whether other measures can better measure the quality of work. Jones responded that bibliometrics is a useful metric because it can be tracked across many documents and is readily available. Jones also clarified that although this proxy may not measure quality, it can measure impact. Work that is cited is likely to be more influential, even if its quality is lacking, however, Jones agreed with Rafols that bibliometrics alone is imperfect and other measures are necessary. Having multiple ways of considering impact will ultimately be more useful.

7

Measurement Efforts

The workshop included several presentations focused on existing efforts to measure convergence. Julia Lane (New York University) reminded workshop participants of the National Science Foundation's (NSF's) description of convergence as using many approaches to address complex problems focused on a societal need. She argued that many societal needs are empirical. To address the needs of the National Center for Science and Engineering Statistics (NCSES), she asked: (1) how does one obtain the data required to answer a given research question, and (2) how will those data be valued after analysis? Lane argued that several technologies could be used to construct datasets, particularly federally funded datasets. As a labor economist, she previously worked with confidential micro datasets, and in her view, one major challenge has been that the current data infrastructure and traditional measures (unemployment, gross domestic product) have become less salient since their initial development. Data infrastructure will need to adapt as needs change. In the past two decades, for example, privacy concerns and level of willingness to provide personal data has drastically changed.

Convergence has led to the availability of new types of data, which can be leveraged to solve complex societal problems, but making data openly available does not ensure that its quality is high enough to be usable. This problem is exemplified by data concerning COVID-19. Massive amounts of data have been collected and are now available. Still, there is little organizational structure to characterize the types of data available, and there is no method to identify who is using the data and whether particular results have been replicated or reproduced. She cited Google Scholar as an example of an infrastructure developed to help researchers find publications. Unfortunately, there is no such infrastructure to facilitate a search for datasets.

Lane said that it is important to consider different ways of collaborating with other researchers while utilizing confidential datasets. New technologies such as FedRAMP allow data to be securely shared via cloud storage. FedRAMP provides a secure data environment that will host, secure, monitor, and control data access, thereby reducing much of the burden associated with using a confidential dataset.

Data science is becoming increasingly important across the United States, and more universities are offering degrees in this discipline. Lane suggested that a shift in how data scientists are trained could be beneficial: Instead of students learning to code "Hello, world" in perfectly curated datasets, they could be trained to work with real-world government data. This approach would allow skill improvement while serving the public good. To facilitate the service-learning idea, it would be important to build an infrastructure that can support various new ideas and approaches.

A greater focus on solving large-scale social problems could begin with large government datasets but later be extended to other data sources. Government datasets account for nearly one-quarter of economic activity and can be used to answer questions related to a variety of topics. Lane's research group has been partnering with a subset of agencies, including the U.S. Department of Agriculture (USDA), National Oceanic and Atmospheric Administration (NOAA), National Institute of Standards and Technology (NIST), and National Institutes of Health (NIH) to determine which datasets should be curated and highlighted. NOAA, for example, needed a means of identifying forests with high conservation value, so it developed metrics based on the biological diversity of ecosystems and how the forest serves community needs. Tyler Christenson (NOAA) has proposed that those measures could be used by the National Forest Service to identify forests with high conservation values.

As examples of such metrics, Lane explained that one could simply count the number of times a dataset is used in a way similar to that used to track publication citations. Such tracking could help identify high-impact datasets by, for example, documenting how datasets are applied to community needs. Through two open-source competitions, Lane's team challenged the computer science community to develop a machine learning and natural language processing algorithm that could be used to identify datasets used in publications (in other words, to create what Lane's team calls a "dataset publication dyad").

Lane said a group of computer scientists built a training corpus and artificial intelligence (AI) model that correctly identified a dataset with 78 percent accuracy. Their next goal is to improve the capabilities by sharing the critical core of dataset publication links with the public research community, which can then help identify mistakes and suggest improvements. Now that proof of concept has been completed, the next stage is community engagement. Lane described the work she has done with the Agricultural Research Management Survey (ARMS), which began with a set of metrics that includes publication topics, publications that cite ARMS, the researchers involved, institutional representation, and the ecosystem where operations occurred. When applied, these metrics begin to answer questions about high conservation value for agricultural research.

Work has also expanded the corpus for CHORUS, a publication and compliance data service. Lane noted that it took six graduate students a few months to build this hand-curated corpus. A machine learning model was then applied to it, which increased the number of publications identified by 30 to 50 percent within 2 hours, depending on the measure. A validation tool was included to verify the results (the datasets used in each publication, that is).

Lane described her work with Mike Stebbins using Kaggle, a data modeling and analysis platform designed for competitions. Kaggle can be leveraged to increase engagement and allows community-based improvements through what is known as the Kaggle process. The process is designed to encourage thousands of computer scientists to improve machine learning models that generate the corpus of publications, with the hope of reaching 95 percent accuracy. Lane and Stebbins hope that other scientists and researchers will engage, improve, and automate the corpus continuously to build an effective infrastructure around database searches.

Federal statistics, federal measures, and the larger research community can overcome the current challenges faced. She believes that this work provides a classic example of convergence: data from multiple sources could be combined in order to answer complex social challenges.

Bethany Laursen (Laursen Evaluation & Design and Michigan State University) shared findings from a study conducted by her private evaluation consultancy and coauthors Nicole Motzer (National Socio-Environmental Synthesis Center or SESYNC) and Kelly Anderson (University of Maryland).

Laursen presented a systematic review of pathways for assessing interdisciplinarity. She began by defining each of the core terms in that effort. First, the review accepts the definition of interdisciplinarity adopted by each author under review: A publication's use of the term "interdisciplinarity" qualified that publication for inclusion. Publications that relied on certain elements of transdisciplinarity but traced the overlap between these elements and interdisciplinarity were also included. The study considers convergence to be a type of interdisciplinarity or transdisciplinarity. Second, the study defines assessment as an empirical summary of important characteristics, with "importance" determined by authors themselves. Finally, the study uses "measure" to include qualitative and quantitative observations.

Laursen suggested that every assessment unfolds along an "assessment pathway," a series of choices that guide assessment efforts. The pathway may include all of the following phases: examining a particular context and the goals within it; developing the intended output of the empirical study; developing criteria and (in a full evaluation) developing standards for those criteria; choosing units of observation (such as temporal or spatial), measures or indicators of those units, and analytic methods to observe those indicators; developing an aggregation method for multiple indicators; and (in an evaluation of quality or merit) developing evaluative judgment methods that can generate a declaration of the object's potential, current, or total quality or merit. Such a declaration could inform a decision to retain or discontinue the object of the assessment. Laursen demonstrated this structure's application by showing two examples of assessment pathways used in an evaluation of the diversity of research teams funded by the National Institute of Aging.

The assessment pathway shows that NCSES researchers must make many choices when assessing interdisciplinarity or convergence and that failure to consider or act on various choices along the pathway could mean an incomplete evaluation. When Laursen and colleagues began their own assessment of interdisciplinarity, they confronted an unmanageable range of methods and pathways to conduct an evaluation of "interdisciplinary synthesis" at SESYNC. They thus shifted their work to a systematic review of all published pathways in the interdisciplinary assessment literature between 2000 and 2019. Using a systematic review design, search criteria, and specific databases, Laursen and her colleagues found 1,006 unique assessment pathways.

Laursen shared four key lessons. First, authors conducting an assessment should either monitor or evaluate and explicitly declare their choice. Monitoring describes measures or observations without making a value judgment about the results. By contrast, evaluation arrives at a value judgment or diagnosis. Authors should choose whether to monitor or evaluate based on whether or not they want to arrive at an evaluative judgment that can warrant further action.

Second, researchers should use rigorous evaluative reasoning to avoid logical "dead ends." Laursen explained that to move logically from a description of a measured criterion to a diagnostic value judgment of that criterion, a researcher needs an explicit standard or scale of value. Moreover, each of these components must be explicit, although at least 83 percent of pathways surveyed by Laursen and team that aimed to form evaluative judgments did not include these minimum elements of rigorous evaluative reasoning. Most thus failed to reach an evaluative judgment and, in the process, did not offer readers a clear picture of how to adopt the pathway to their own assessment efforts.

Third, researchers should use mixed methodologies to observe multiple facets of interdisciplinarity. Of the studies identified by Laursen and her coauthors, 81 percent used exclusively qualitative or exclusively quantitative methods. Only 10 percent used mixed methods. Among the studies that employed mixed methods, many did not describe how multiple measures were aggregated to support their final judgments. Of all the pathways identified, moreover, 6 percent used only one measure to form evaluative judgments, which seems inadequate to measure the highly multifaceted phenomenon of interdisciplinarity. Laursen highlighted rubrics as an exemplar method for evaluating multidimensional, multifaceted, and complex phenomena such as interdisciplinarity and convergence. A rubric allows a researcher to specify standard performance levels for each criterion. As Laursen explained, the rubric must specify criteria that are accurately measured and define appropriate, detectable performance levels.

Fourth, Laursen encouraged researchers to use the dataset of pathways developed by her team to identify criteria, standards, measures, methods, and research approaches that may be appropriate for assessing convergence. Laursen and colleagues have built an interactive dashboard that allows the user to filter and show dominant modes and approaches that have previously been used to assess interdisciplinarity.[1]

Laursen closed by arguing that the published literature on assessing interdisciplinarity generally lacks evaluation basics. Well-established basics will help to increase the rigor of assessments and provide a shared vocabulary among practitioners. Practitioners should choose, explain, and follow their assessment pathway clearly and carefully.

Dan McFarland (Stanford University) discussed his work deriving convergence measures using network analysis and natural language processing. He focused on social convergence (how stakeholders relate to one another in intellectual creation) and intellectual convergence (how conceptual relations and understandings arise from scientific discovery). McFarland mapped these dimensions of convergence onto two axes that converge in various forms of interdisciplinary engagement, from macrolevel "trading zones" across disciplines within universities to shared projects at interdisciplinary centers, the development of "boundary objects" shared by multiple fields (such as a patent), and conceptual linkages studied by the philosophy of science.

McFarland has focused on investigating the impact of interdisciplinary centers on social convergence by answering three key questions: Do they generate productivity and new collaborations? Do they hinder high-quality disciplinary work? Do they induce unique forms of social convergence? McFarland explored these questions by examining the impact of several

---

[1] See https://shiny.sesync.org/apps/evaluation-sankey.

Stanford University institutes on departmental connections and the composition of graduate dissertation committees. The first institute is BioX, an interdisciplinary bioengineering center the founding of which strengthened rather than disrupted preexisting collaborations and created a kind of "thoroughfare" that facilitates activity without disadvantaging individual departments. In another case, the Woods Environmental Center generated a new intellectual community on campus that has led the university to consider creating an entirely new degree-granting school.

Intellectual convergence, the exploration of the relationships between ideas, has also been studied. Bibliometric analyses and citation practices have been the typical proxy for this type of convergence. McFarland has found that citing work outside of one's discipline can be a high-risk practice (although there is also a corresponding potential for high reward). McFarland also considered the impact of annual review papers as a form of synthesizing convergence. He found that authors mentioned in annual reviews actually experienced a decrease in citations of their own work, as the review was increasingly cited in place of the papers it discussed. McFarland described this phenomenon as a type of creative destruction that suggests convergence may be symbolically violent: It requires eliminating linkages in the interest of simplification and creating new methods to bridge and integrate.

McFarland has explored potential links between intellectual and social convergence such as including the directionality of this relationship (whether social connections promote intellectual convergence or intellectual interests foster social convergence), as well as factors that facilitate both forms of convergence. McFarland finds that social convergence frequently crosses ranks and ages (senior and junior faculty are more likely to collaborate with each other) and that cultures of collaboration differ widely between STEM fields (which tend to follow a more rapid pace of team scientific discovery) and non-STEM fields. Changes also emerge over time: With the development of the Internet, intellectual connections tend to precede and lead to social collaborations, while the opposite order tended to unfold before the Internet existed.

This directional shift led McFarland to ask whether the drivers of convergence are primarily intellectual or social, but he eventually concluded that these factors are often simply distinct in various research products: Collaborating researchers from different fields, for example, may adopt a division of intellectual labor that preserves their disciplinary differences such as writing discrete sections of a publication. Truly mixed-method research may treat a particular discipline as the primary one rather than stage an interdisciplinary encounter of equals. McFarland acknowledged that the aspirational model of a true merger of disciplines does exist, typically in the form of a graduate student or postdoc who synthesizes expertise in multiple disciplines. Such researchers, though, typically do not consider the implications of convergence for their disciplines but instead focus more narrowly on specific publications or products.

McFarland concluded by discussing examples of intellectual convergence. Using the field of natural language processing as an example, McFarland showed that certain fields have "phenomena anchors," or core problems that remain remarkably stable over time while the methodologies to study them are much more volatile and continually changing. Other fields, however—such as biology—may see continual change in the phenomena they study and not just the methods they use to do so. McFarland suggested that intellectual convergence across such disciplines would involve finding a way to operate across the levels of phenomena and method.

In his talk, Ismael Rafols (University of Leiden) argued that gathering different types of knowledge to foster innovation and address social challenges requires consideration of two issues: epistemic integration (how knowledge is combined) and the direction in which knowledge combination moves. Using COVID-19 as an example, Rafols said that knowledge from various disciplines could be combined toward risk communication, vaccine development, health care management, and more. The description of directions of research requires new kinds of multidimensional metrics that describe portfolios of research and new ways of communicating those portfolios (including potentially tables, networks, and maps).

Previous efforts to develop universal or general indicators of interdisciplinarity for policy use has been unsuccessful. Rafols, for example, served as an advisor to efforts by the British Council performed by Elsevier and then by Digital Science to develop national measures of interdisciplinarity that generated inconsistent, conflicting, and confounding results. Indicators of interdisciplinarity developed in previous studies had been successful and valuable for particular research areas but did not apply to the variety of contexts at the national level. Wang and Schneider have recently reported similar findings regarding inconsistency of universal indicators.[2] Rafols concluded that using a single or even a few atomistic indicators to measure complex research activities capable of addressing societal problems is misguided.

Rafols proposed shifting from an atomistic to a portfolio approach modeled on ecology. An ecology portfolio would not map degrees of interdisciplinarity: It would instead investigate the entire landscape that makes convergence possible. As with environmental ecology, it would connect structure to function by relating the production of science to an organization's mission or societal need. Descriptions of the portfolio landscape would include the variety of knowledge in the landscape rather than within

[2]Wang, Q., and J.W. Schneider. (2020). Consistency and validity of interdisciplinarity measures. *Quantitative Science Studies* 1(1):239–263.

individual papers, diversity in the social background of researchers (which Rafols likened to soil conditions), linkages between bodies of knowledge in the landscape, and the balance and directions of knowledge. Shifting to a portfolio approach avoids unidimensional indicators in favor of a multidimensional map in which different directions are possible. The new framework contains both the distribution of knowledge ("spaces") and efforts to integrate that knowledge ("bridges"). Bridges may exist at the organizational, cognitive, or societal levels.

Rafols offered an example of landscape mapping at the organizational level.[3] Using a global map of different scientific fields as a starting point, he mapped the diversity of scholarship in departments at the London Business School and the Institute for the Study of Science, Technology, and Innovation (ISSTI) in Edinburgh. Linkages across fields can be investigated using bibliometric data to determine whether expectations of crossover are met or exceeded. Rafols' approach allows one to see not only the fact of interdisciplinarity but also the directions in which it is proceeding, which cannot be determined from unidimensional measures of interdisciplinarity.

He moved on to show convergence as a combination of topic distribution and integrative efforts. The combination allows for a variety of meaningful perspectives that are flexible to the question asked. Rafols explained further with a mapping of rice research[4] that represented six distinct topic areas and tracked efforts in those areas over time to understand how research has evolved. This framework also allows one to visualize how research priorities may differ across countries that have distinct needs and concerns.

The kinds of analyses Rafols presented illustrate the characteristics of mapping a convergence landscape. This approach measures types and directions of knowledge combinations, not just the quantity of combinations that are captured by interdisciplanarity indices. This approach also relies on maps that can visualize the distribution of research and linkages between topic areas rather than traditional statistical reports. As previously mentioned, Rafols suggests a combination of top-down and bottom-up approaches.

Finally, it is vital to consider what type of convergence is most relevant. Measurements could focus on changes in organizations such as university departments or in relation to specific societal problems.

Josh Schnell (Clarivate) followed with his presentation on the value of bibliometric approaches. He acknowledged that multidimensional value

---

[3]Rafols, I., L. Leydesdorff, A. O'Hare, P. Nightingale, and A. Stirling. (2012). How journal rankings can suppress interdisciplinary research: A comparison between innovation studies and business and management. *Research Policy 41*(7):1262–1282.

[4]Ciarli, T., and I. Rafols. (2019). The relation between research priorities and societal demands: The case of rice. *Research Policy* 48(4):949–967.

systems of science are not always well reflected in science metrics. New measurement efforts ultimately will depend on what NCSES and its stakeholders want to know about convergence within a national context. Schnell described one attempt to map the trajectory of research approaches employed by Japanese researchers. The study[5] found that approaches became less diverse and more conservative over time. This mapping analysis was then used to reveal to policy makers, university leaders, and researchers how science was actually being practiced.

Convergence can be viewed from either a theoretical and conceptual perspective (the science of how convergence should be studied) or from a practical perspective (the question of how researchers are actually collaborating). Any study of convergence involves multiple dimensions, including disciplinary interactions and topics; the problem-orientedness of the research; people and organizations involved; institutional context; tools, equipment, and methods involved; the learning and educational dimension; and infrastructure. Schnell added that convergence efforts do not all need to be novel: A convergence lens can be applied to existing NCSES data, and linked datasets can be particularly important for measurement. Interdisciplinary data on the Survey of Earned Doctorates (SED), for example, could be linked to bibliometric data to offer a view of both processes and product. Although challenging and time-consuming, the linkage of datasets could enable researchers to ask increasingly exciting questions and reflect more on multidimensionality.

Schnell discussed several examples that demonstrate the value and limitations of bibliometrics. He shared a map of emerging highly cited research, or "research fronts," that visualized disciplinary and topical interactions. This kind of mapping can be based on bibliometric data, however, bibliometric analysis works best when it is refined and linked to other data. Measurements of the role of researchers and organizations in convergence can, for example, be refined by considering data on the impacts of COVID-19 on women in the research community. The field similarly still lacks a proxy measure of problem orientation that is based on bibliometric data, and bibliometric analysis itself has limitations for national-level interdisciplinary research (IDR) or convergence (as Rafols' presentation also noted). In addition to refinement of these technical challenges, Schnell suggested that bibliometric analysis should be informed by and confirmed with stakeholders.

Potential for the further use of bibliometrics hinges on a more robust understanding of the theory and practice of convergence, which Schnell's team at Clarivate is working to produce. Surveys and field research should be used to understand several elements of convergence, including tools

[5]Igami, M., and A. Saka. (2016). Decreasing diversity in Japanese science, evidence from in-depth analyses of science maps. *Scientometrics* 106:383–403. doi:10.1007/s11192-015-1648-9.

and equipment needed, methods involved, and learning and educational dimensions. Schnell sees a potential in bibliometric analysis for considering citation networks, to examining the process of convergence through text-mining techniques, and analyzing the full text of scientific articles to understand citation context. These mining techniques hold the potential to visualize the integration of knowledge as it pertains to convergence.

Michèle Lamont (Harvard University) spoke about a socioemotional cognitive platform for interdisciplinary collaboration based on research conducted in 2010 and published in a 2016 paper with Veronica Boix Mansilla and Kyoko Sato. Their study is based on in-depth interviews conducted with individuals from nine different research groups—supported by grants from the Canadian Institute for Advanced Research (CIFAR), the MacArthur Foundation, and the Santa Fe Institute—to understand how they define successful interdisciplinary collaboration. The interviews focused on how members of the group define markers and facilitators of success; how they experience the construction of group membership and collective work; and the research infrastructure created by funding organizations that enable success.

Interviewed participants were life and social scientists who had collaborated for 1 to 8 years. In addition to conducting interviews, Lamont's team also observed meetings, administered questionnaires, and reviewed the research team's publications. To capture the respondents' multidimensional experiences, Lamont's team proposed a shared socioemotional-cognitive (SSEC) model that registers not only the cognitive dimensions of interdisciplinary collaborations—which are generally addressed in the extant literature—but also its emotional and interactional dimensions.

The SSEC model highlights the importance of intellectual, interactional, and institutional factors. Interactional factors include emotion, which plays a constitutive role in creating and sustaining successful interdisciplinary collaboration. Collective excitement was mentioned by 51 percent of respondents; 28 percent described the joy of collaboration; 53 percent mentioned the importance of group deliberation and learning how to learn from each other; and 32 percent noted the value of developing meaningful relationships. Within intellectual or cognitive factors, 67 percent mentioned the importance of cross-disciplinary exchange, and 40 percent mentioned sharing intellectual tools. Various factors facilitate these valued activities, such as socializing outside of meetings. Each of the nine groups experienced cognitive, emotional, and interactional factors in different proportions, with cognitive factors the most salient.

The intellectual dimension of interdisciplinary collaboration is marked by several features. First, group members participate in productive problem framing: The definition of their research problem evolves iteratively over time as group members engage in codefinition. An optimal level of ambiguity is essential to this process. Having an overly strict definition of a problem may impede interdisciplinary collaboration, while the appropriate amount of ambiguity will allow the group flexibility and provide a means to develop the interdisciplinary exchange. One participant suggested that ambiguity allows every creator to offer their expertise and analogized the process to building scaffolding as work progresses.

Lamont stated that a full understanding of interdisciplinary collaboration is impossible until attention is paid to emotional aspects. One participant highlighted this importance, comparing interdisciplinary research collaborations to marriage. Emotional success was enabled by feelings of group belonging, respect and admiration of peers, a climate of conviviality, and effective leadership by individuals who understood the cognitive, emotional, and social demands of successful interdisciplinary collaboration. Although many remarks emphasized the positives of interdisciplinary research, Lamont and colleagues found that participants did mention markers of unsuccessful collaboration, including the failure to frame a problem in ways that are shared by network participants and the failure to establish relatively shared expectations. They also pointed to several barriers to successful collaboration, such as the self-interestedness of individual members, disciplinary closed-mindedness, disciplinary language, and conflicting ideologies, epistemologies, and communication styles.

DISCUSSION

Nora Cate Schaeffer (University of Wisconsin–Madison) asked Laursen whether studies using multiple measures linked each measure to a conceptual definition. Laursen confirmed that links were nearly always made to a conceptual definition. Publication authors often named a construct or subconstruct, and explicitly explained how they proposed a new measure of that construct.

Klein and Kara Hall (National Cancer Institute) both asked Lamont about the potential implications of NSF's effort to measure socioemotional and cognitive aspects of convergence. Lamont noted that of the three organizations she studied, CIFAR does not have a short-term deliverable. She also noted that open-endedness appeared to allow researchers to be more adventurous in their collaborations. She thus suggested that this fact might lead to the creation of funding models that are less explicitly tied to proposals with well-defined, expected results. She also acknowledged that this type of research funding is expensive and likely to meet resistance. She suggested, nonetheless, that a more inductive and multidimensional focus on emotions and relationship creation in interdisciplinary work might increase alternative forms of collaboration and funding outside of NSF.

McFarland was asked how network analysis could help assess progress made on reducing barriers to convergence. He explained that concept networks are used to identify the locations of problems and bottlenecks. His research investigates where connections are made or lost and identifies pervasive problems or gaps, and based on that research, he concluded that network analysis cannot determine whether filling a gap is truly a sustainable solution.

Hall followed by asking McFarland how he might conceptualize measurements that could examine interdisciplinary and emergent research on a national scale. McFarland responded that the objects of network analysis frequently require smaller networks. Some methodologies can be used for more extensive networks but they are computationally taxing. Additionally, it is possible to "community detect"—discover cohesive groups or clusters—within large networks and then perform more computational procedures on the resulting data. McFarland is currently drafting a textbook for social network analysis in order to address analysis at varying scales.

Hall asked Laursen whether she could share any promising measures of interdisciplinarity that NCSES could use to explore convergence. Laursen noted that her team extracted 890 unique measures from 142 publications investigated. Laursen said that the most promising measures were combinations or aggregations of quantitative and qualitative measures because they unite a countable output (such as bibliometric data) that measures the products of convergence and a qualitative measure that helps unpack the process of convergence. Qualitative measures are generally not as clearly defined as the quantitative measure, but dozens of incisive critical questions are available to investigate the process aspects of convergence. These questions often still have several weaknesses: They almost always take a yes/no format that is difficult to translate into indicators for action. They also often implicitly embed standards in a way that obscures what counts as a "good" answer, and publications often do not explain how to synthesize the answers to a list of questions in order to derive a coherent overall judgment. She added that the method used to aggregate measures can be even more important than the measure itself because it is closer to the final judgment of quality. Researchers who use a rubric to aggregate measures can synthesize multiple qualitative and quantitative measures in a clear way. Coauthor Nicole Motzer (SESYNC) added that the group's upcoming work will include a cluster analysis that investigates which published works move furthest along their assessment pathway. That analysis may provide guidance for convergence measures. Laursen noted that the group also plans to publish a subanalysis of scaling differences.

Khargonekar asked Rafols to describe the distinction between direction and combinations of knowledge. Rafols said that combinations bring different bodies of knowledge from various disciplines together, while direction captures how these types of knowledge lead toward specific research agendas and influence the kinds of innovations that are possible.

Schnell was asked how his company, Clarivate, approaches coverage and disambiguation of individuals in bibliometric analysis. He responded that both are areas of concern for Clarivate. Disambiguation has been aided by advancements in machine learning technologies and by ORCID, which creates a unique identifier for authors, but more work remains to be done. Schnell noted that complete coverage is impossible and thus reiterated the importance of aligning decisions about research on convergence with a specific purpose. Researchers must determine what kinds of literature and topic areas should be measured and which aspects of convergence they seek to understand. Schnell also addressed concerns about bias within bibliometric data, noting that work has been done to understand potential biases and reiterating that bibliometric analysis may leave gaps that require different dimensions of measurement to fill.

Both Rafols and Schnell were asked about the prospects for providing interactive maps and data visualizations that would be useful to various stakeholders. Schnell stated that interactive tools could help policy makers and the public understand various policy questions better. He cautioned that these tools are not always comprehensible for lay audiences, so it might be good to use other methods to convey information about convergence or a portfolio of research. Rafols agreed and pointed to other limitations on interactive mapping. He requested input from those with expertise in qualitative research to assist in developing qualitative processes to enhance productive use of interactive quantitative tools.

The two presenters were asked about the use of bibliometric analyses of convergence in forecasting emerging trends or fields. Schnell warned that the term "prediction" can be problematic, but agreed that bibliometrics can provide some indication of emerging trends and fields. On the other hand, he noted that the datasets of bibliometric analyses are produced on a delayed time frame (there is a lag between completion of a work and its eventual publication). Rafols emphasized the need for caution in using the term "prediction." He suggested that because convergence always moves in the direction of some research agenda, researchers can use interactive tools to consider "what futures we want to build" rather than to predict what future will inevitably come to pass.

The moderator asked Rafols to elaborate on the question he raised on the first day of the workshop about the use of networks and scientific mapping to measure convergence. Rafols explained that he is interested in new kinds of indicators that consider cognitive distances between the categories mapped in studies of convergence. Distance is important because when subject areas are more distant from each other (sociology and biochemistry, for instance), it becomes increasingly difficult for the involved researchers to understand one another.

Khargonekar asked the two presenters whether it would be useful to distinguish between predefined intentional trajectories or vectors of convergence and serendipitous or emergent trajectories or vectors. Schnell answered that more research is needed to understand the degree of directedness of convergence in practice, both in NSF grants and in the scientific literature. He also emphasized the importance of ensuring that these practices are captured in bibliometric analyses. Rafols suggested that intentional and serendipitous trajectories do not always exist in dichotomous form. He presented the example of the Zika virus vaccine, which was serendipitously developed based on research from related areas such as dengue fever. Mapping, however, would allow a user to anticipate the areas from which such serendipitous connections can most easily be—or are more likely to be—made.

The final questioner noted that a problem orientation is frequently the motivation for convergence research and asked whether it is also a necessary condition. From the perspective of Clarivate, Schnell stated that convergence does require a specific or compelling problem to be its driver. Although nonconvergent research can be problem-oriented, convergence is distinguished by combining deep integration across disciplines with a necessary orientation toward solving a specific or compelling problem. Rafols addressed the question from the other direction: He noted that convergence is a highly desirable if not necessary feature of robust, problem-oriented research. Excessively narrow approaches to problem-oriented research, he suggested, are often the flaw responsible for poor-quality policy advice, since they may not consider all relevant angles.

8

Feasibility and Implementation

Several prominent survey methodologists participated in a discussion on the feasibility and considerations for potentially implementing a national-level data collection on convergence. Graham Kalton (Westat) focused his comments on two areas of interest to the committee: (1) Relative to other concepts that we can measure in the context of existing science and technology indicators, and within current National Center for Science and Engineering Statistics (NCSES) surveys and reports, how valuable would it be to construct national-level measures of convergence or interdisciplinary research (IDR), and should these measures be sought in addition to existing measures or should they replace some existing measures? (2) What if any kinds of questions about the emergence of convergence/IDR are of greatest value to the nation and, in particular, which questions provide valid and actionable information?

Based on the workshop discussion to that point, Kalton observed that although the multidimensional aspects of convergence seem to be essential, they also create real challenges with respect to measurement in existing large-scale surveys. If one or two dimensions could be identified that could be used to provide imperfect indicators of convergence, though, it might be possible to collect data on a national scale that could be used to move the field forward. Because resources are limited, Kalton suggested that National Center for Education Statistics (NCES) might collaborate with other agencies such as NCES or the Bureau of Labor Statistics. He referenced the presentation by Anderson on the Business and Enterprise Research and Development (BERD) and Higher Education Research and Development (HERD) surveys. It also occurred to Kalton that another data collection vehicle that might be amenable to some modification is the Survey of Doctorate Recipients (SDR), a panel survey that could be tapped to collect longitudinal information about activity in convergence research among doctorate recipients. Kalton acknowledged that the feasibility of producing one or two indicators must be addressed first. He wondered whether there was any evidence that another country has successfully applied this approach at a national level.

Kalton expressed concern about the quality of data extraction by non-specialists to answer complex questions that incorporate cross-references to lengthy definitions of terms. He was skeptical that the data extractors would fully comprehend the definitions and then reliably search out the data, any failure to do so undermining the quality of the responses. Kalton thinks that the quality of the survey responses to complex questions on convergence needs to be carefully evaluated.

If one or two measures related to convergence were identified as useful indicators of the developments in this field, then this would raise questions about how they could be used to guide actions by the National Science Foundation (NSF) and others. Knowing which combinations of fields of expertise and in what settings convergence is more likely to occur would be valuable for targeting research programs. Kalton, however, remained pessimistic about the feasibility of identifying a small number of questions or concepts that could satisfy the needs of NCSES at this time. Although there is a high level of interest in convergent research, difficulty arises when applying the concept on a national scale.

Adopting an approach rooted in the traditions of the classical social sciences, Nora Cate Schaeffer (University of Wisconsin–Madison) began her remarks by asking what are the research questions to be answered, which is another way of asking how measures of convergence are to be used, and what is important for those measures to capture. For NCSES, it also means placing the definition and measurement of convergence within the context of the research enterprise. She followed with a number of other questions a social scientist routinely asks in the face of a measurement challenge: What is the underlying causal model of the construct? What units are involved in the causal process or causal processes? What characteristics of those units are important to measure, what units can be used for collecting data, and are the units from which data are collected and those for which measurements are needed different? Concerning the challenges presented in this workshop, Schaeffer asked, what unit of analysis does convergence belong to, what is its conceptual definition, and what is its operational definition?

To help explicate this approach more systematically, Schaeffer presented a tentative conceptual model of the social processes and actors that generate convergence (or not) (see Figure 8-1, showing units in yellow text, with some characteristics of those units noted in italics). This model depicts convergence as a property of research projects produced by researchers

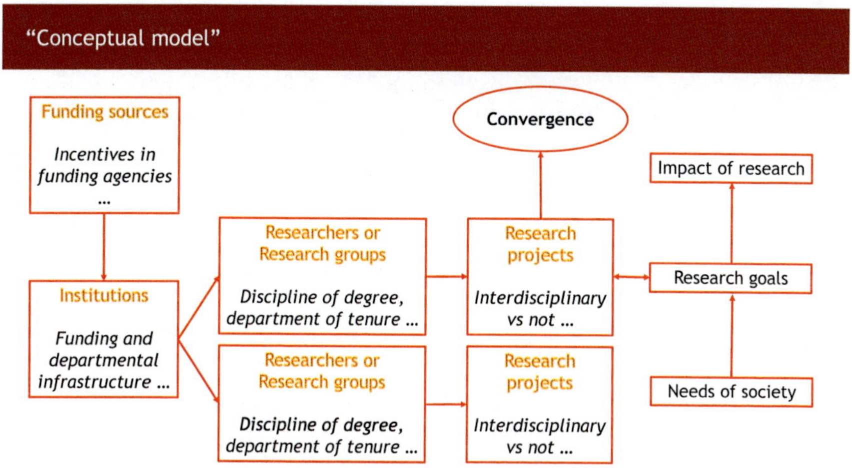

FIGURE 8-1 Tentative "conceptual model" for measuring convergence.
SOURCE: From Schaeffer (2020) Presentation at the Workshop on the Implications of Convergence for How the National Center for Science and Engineering Statistics Measures the Science and Engineering Workforce, October 23. Available: https://www.nationalacademies.org/event/10-22-2020/a-workshop-on-the-implications-of-convergence-for-how-the-national-center-for-science-and-engineering-statistics-measures-the-science-and-engineering-workforce.

within departments that are housed within institutions to which agencies make grants, although for some purposes, convergence might be conceptualized as a property of research teams or grants. The diagram is intended to illustrate some implications of clarifying how convergence is defined: To what unit should convergence be attached? Is convergence a property of a project? Is it a property of the goals of the project? Many different possibilities exist, and it is probably premature to settle on just one possibility. She added that the definition of convergence and the causal, social, and scientific processes underlying convergence may change as a field evolves, ages, and spreads around the world.

Schaeffer noted that funding agencies and their calls for applications could create incentives for institutions and researchers to pursue convergent research. Institutions are critical to this effort because they supply the infrastructure needed to conduct convergent research and the incentives to do so. The workshop participants heard absorbing discussions about the ways in which universities can foster interdisciplinary or convergent work, such as making bookkeeping and accounting systems communicate seamlessly across different units. These institutions are also the data collection units for the NCSES surveys that serve important roles in the measurement of convergence. It therefore can be problematic if convergence is defined only at the level of projects, papers, discoveries, or patents but left undefined at the institutional or university level, especially when thinking about options available for implementing a definition or a measurement strategy. Schaeffer does not see a path toward a "unitary" definition of convergence that is relevant for all research goals or all units, but also does not consider a single definition to be necessary. Instead, she sees convergence as a concept with different meanings, different units, and activity at different points in time. If the measurement strategy remains flexible, then lack of a single definition for convergence is not necessarily a problem, as long as investigators specify which conceptual and operational definitions they are using.

Schaeffer mentioned three such concepts that are closely related to but distinct from the concept of convergence: interdisciplinarity, collaboration, and synthesis. If these are seen as dimensions of convergence, and each dimension is measured separately, the resulting data might be reliable and flexible, which could make it easier to monitor change over time. A single concept of convergence that is broad and vague is not necessarily plausible or necessary. If convergence is a property of projects, then measuring convergence at the level of the project may be reasonable. If, for example, researchers focus on interdisciplinarity, which seems to be one dimension of convergence, an operational definition that uses the standard departmental designations of the researchers on a project could be implemented at the level of projects. Information gathered for projects could be collected by institutions and aggregated to yield measures at a national level. Such measures could be more reliable than measures that ask investigators to assess the convergent qualities of their research projects.

She posed the possibility of measuring convergence at the level of the grant award. If funding agencies can agree on a definition of convergence and tag characteristics of a grant, then collecting and analyzing grant-level data should be feasible and potentially reliable. Another idea would be to continue the relationship with universities and support them in the development of infrastructure that allows for more straightforward implementation of measures of the critical concepts.

Andrew Zukerberg (National Center for Education Statistics) began his comments by focusing on the task of operationalizing convergence and its measurement. He highlighted the impact of convergence and the measurement of its impact on science and scientists. He asked whether investigations should have an indicator that serves as a baseline frequency that can be used as a benchmark for comparisons or a measure of the impact of the convergence. An indicator could focus on the frequency of convergent projects or the level of funding directed to convergent efforts. Zukerberg noted that impact was frequently mentioned and therefore should also be defined.

Zukerberg reviewed the definitions of convergence, beginning with the one provided by the National Academies in 2014.[1] He highlighted two primary characteristics within this definition: deep integration across disciplines and research that is driven by a specific and compelling problem. Adding a component that addresses improving quality of life would appropriately reflect the comments heard throughout the workshop. He agreed with Kalton and Schaeffer about the challenges of measuring an indicator that lacks a standardized definition. Still, he continued, unidimensional characteristics are easier to measure without a standardized definition in place. He acknowledged the issues inherent in providing a definition within a survey instrument. Even if an appropriate definition is included with the questionnaire, one cannot assume that the respondents will read the full definition and answer in the intended fashion.

Zukerberg offered a workaround to the definition issue using his research on bullying (which is based on Bureau of Justice Statistics data) as an example. When a concept such as bullying or stalking is presented, participants have immediate reactions. If instead individual components of a larger concept are presented, a scale can be built from the individual components in place of a standardized definition. For bullying, a standardized set of individual concepts might include answers to questions such as: Did someone exclude you intentionally? Did that person have more power over you? Did the event happen repeatedly? Along similar lines, if a standardized definition of convergence is not possible, measuring standardized concepts that make up aspects of convergence such as translational ability, diversity in disciplines, and a specific or compelling problem could help. Although it can be burdensome to divide many different concepts into parts, the individual components can be used to build a complete measure.

Zukerberg emphasized the need to consider the opportunity costs when adding a measure, which may require the removal of other questions from a study or replacing entire studies. One must consider whether the data provided by the new measures are valuable enough to justify the deletion of currently included scales and measures. Still, issues may arise in retrieving a complete dataset, particularly for an outcome aspect of research, because of the complexity of the underlying constructs. To understand questions regarding a larger-scale or long-term impact on researchers, NCSES could conduct a longitudinal or panel study (such as the Survey of Doctorate Recipients, or SDR) with measures based on its priorities.

[1]National Research Council. (2014). *Convergence: Facilitating Transdisciplinary Integration of Life Sciences, Physical Sciences, Engineering, and Beyond.* Washington, DC: The National Academies Press.

DISCUSSION

After the formal remarks, Jolene Smyth posed the question of whether attendees at the workshop think that it might be possible to identify one or two dimensions of convergence with a level of agreement that would warrant commencement. Jason Owen-Smith replied that efforts should be made to begin investigations, even if a perfect strategy has not been laid out. He believes that there should be a working definition paired with concrete examples and testing that could be revised in an iterative method, mainly because there are empirical components. Holbrook agreed with Owen-Smith and stated that an iterative experiment would be the best approach to take until consensus is reached on either a single definition or the most salient components. In his view, lacking a consensus should not inhibit efforts. Future progress can be expected to build on the initial efforts, react nimbly, and reverse course if necessary.

Schaeffer asked what unit would be the most useful for measuring convergence and whether the answer that researchers and NCSES give to this question differs. She elaborated that thinking about units would simplify thinking about the dimensions of a concept.

Hall raised the topic of measuring social impacts from the workshop's first day. She added that considerations of social impacts should be carried out on a problem-specific level: identifying the degree of integration and variation of contributors (e.g., experts, patients, community stakeholders) represented in a publication or investigating preregulatory reports that influence policy regulation. An investigation of policy-impacting reports would signal the impact of research on the greater society.

Entwisle revisited Kalton's initial question about why there is a desire to know more about convergence. Reflecting her inclination for a portfolio type of approach, Entwisle recognized that strategic investments in project areas are necessary and are an input to the research portfolio. Less clear are the outputs. In addition, as others have previously mentioned, there is interest to focus on impact as well as outputs. She added that there should be a clear distinction between the units of observation (e.g., funded projects, published articles, and investigators) and units of analysis (e.g., investigators, institutions, and fields) because this distinction will hold considerable implications for the data collection procedures. Smyth added that the determination of what will be the most critical issue to address is the core challenge.

Kalton shared concerns about the societal impact component of convergence ideals. For example, he mentioned the time delay between the conducting of research and the measuring of societal impact. Additionally, he had concerns about deep integration. To identify deep integration, one would need to obtain information from the project director or someone at that level. In his own work as a survey researcher and as a biostatistician, he has often been involved in multidisciplinary research in which a team of researchers from different disciplines have collaborated to conduct a research project. He would not classify this team research as deeply integrated simply because it included researchers from multiple disciplines. He asked how one would measure deep integration and how to exclude the many cases in which the multidisciplinary efforts are standardized and not fully integrated.

Kalton asked if and how interdisciplinarity has been measured at a national level in other countries? Hall noted that in the coming weeks, she would be meeting with a group of scientists from the Swiss equivalent to the NSF at a workshop that would grapple with similar issues of defining and measuring convergent research. She is hopeful that an international effort might allow the countries involved to help one another and learn from each other.

Josh Schnell (Clarivate) had asked the group if determining what convergence is not might be easier than defining what convergence is. Smyth restated Michèle Lamont's (Harvard University) research findings that convergence includes dimensions beyond a cognitive state such as emotional and interactive components. Elements of Schaeffer's statements were also included, and Smyth reminded members of the difference between convergence and collaboration. Determination of the line where collaboration advances into convergence could become an indicator but the overlap between the two needs to be addressed in order to measure the advancement accurately.

Hall stated that there are different degrees of integration that would allow a distinction between terminologies to be made when measuring collaboration or team science. Her thoughts centered around NSF's conceptualization of convergence and the levels of integration that might be expected or included in considerations of convergence. Smyth added that there might be unconventional ways to measure aspects that were previously unrealized as indicators of convergence and that inform a level of interactivity or collaboration occurring in research projects.

Schaeffer observed that interdisciplinarity may be less valid for some of the research purposes for which the concept of convergence is currently being used, but it might be more reliable because operational definitions could be based on an existing institutional infrastructure and a known collection of disciplines or departments. Though the infrastructure provided by a collection of disciplines is imperfect, it is reliable, flexible, and usable. She reminded attendees that emergent fields could become schools and departments within universities relatively quickly, as displayed by Dan McFarland's (Stanford University) research. Using disciplines as a supplemental index could be useful to build a larger concept of convergence, the full definition and explicit measurement of which currently eludes the research community.

Smyth shared comments made in the virtual chat where participants asked how many items could be included for measurement within a survey. Zukerberg stated that the number of inputs varies, particularly based on the context in which they are used. He assumed any measures of convergence would be a component of a larger survey, highlighting the need to balance the complexity of the survey. He added a question about the outcomes aspect and the size of the problem to be solved. Considerations of output will impact the unit of measure. He added that if a funding agency is providing resources, it implies that the problem has been deemed sufficiently necessary and other components should be weighted more heavily.

Erin Leahey (University of Arizona) mentioned Subject Categories in the Web of Science Core Collection—there are currently over 250 of them—and their potential use to advance convergence, particularly over time. According to Leahey, the list of subject categories changes quite regularly, even multiple times per year. There is no longer a single category for psychology, for example: It has been split into multiple types of psychology (developmental, clinical, and social). Changes in the Web of Science Subject Categories reflect areas that are merging, being excluded, or added over time, and this might be able to indicate the nature of convergent activity. Smyth added that the NCSES field-of-study categories for the dissertation question are another avenue for examining how fields have evolved over time.

Owen-Smith expressed surprise that the group's discussion seemed to have quickly defaulted to measuring convergence through a sample survey using a multifactorial question structure. In his opinion, the group is not yet ready for that level of discussion. He asked the group if convergence could be identified on sight but not explicitly defined. Given the lack of a complete definition and in light of the nonsurvey data presented in the workshop, he believes it might be worthwhile to identify projects deemed to reflect successful convergence efforts. New research could follow successful approaches to work toward an empirically based set of metrics before developing a formal definition to be used in a national survey. Smyth commented that perhaps we need a convergent approach to studying convergence.

Hall responded to Owen-Smith's comments and reiterated that the distance between disciplines and the numbers of disciplines involved are critically important. There is, for example, a large difference between a research project that integrates developmental and social psychology with health psychology versus a project involving a social psychologist working with a physicist. The differences in perspective on what constitutes interdisciplinarity will depend on the individual researchers' orientations and the intellectual perspectives they are accustomed to within their own disciplines. Experts working in disciplines and organizational structures that do not tend to collaborate broadly (e.g., due lack of rewards, recognition for cross-disciplinary collaboration via tenure and promotion reviews) are likely to face more barriers and have less experience considering how they might apply their knowledge to a broader problem or communicating how collaborators outside their discipline may integrate their specialized knowledge when addressing common problems.

Smyth built on these points to remark that if a measure is to be included in a survey for humans to answer, it is essential to remember that the respondent's understanding of the topic will inevitably influence how they answer the questions presented. She encouraged the group to consider how people may have different understandings of interdisciplinary research. Smyth noted that it may be premature to consider adding measures to a sample survey because it could have unintended consequences.

9

Summary of Key Themes and Thoughts to Move the Work Forward

Entwisle invited each of the steering committee members to share their thoughts on key themes and on ways of moving convergence measurement efforts forward. The steering committee formulated six discussion questions, and each member was invited to focus on the areas they considered most salient:

- What would be the value in constructing national-level measures of convergence?
- What ideas and strategies would be most worth pursuing to push thinking on this initiative forward?
- What framework could the National Center for Science and Engineering Statistics (NCSES) use for measuring convergence?
- What definitions and criteria would we consider appropriate and what are their limitations?
- What types of NCSES products would be most valuable and for what audiences?
- What should NCSES prioritize as part of its work in this area?

Hall began by reading the National Science Foundation (NSF) mission statement: to "promote the progress of science; to advance the national health, prosperity, and welfare; and to secure the national defense and for other purposes."[1] This expansive mission covers complex societal and scientific challenges, necessitating multilevel, multidimensional, multifactorial,

[1]From the National Science Foundation Act of 1950 (P.L. 81-507).

and multidisciplinary approaches that are similar to the convergence approach. Hall described the evolving context under which research occurs, in which advancements in technology make it possible to analyze problems of greater complexity, more numerous variables, and a greater variety of variables, as well as to enhance the speed, scale, and convenience of interactions and collaborations. While advances in technology over the past century have enabled the development of tools and platforms to support asynchronous and synchronous communication, the COVID-19 pandemic has created an imperative to establish new collaborative norms, including the increased acceptance of virtual meetings for conducting business.

Hall mentioned a number of elements that facilitate intentional and evidence-informed convergence research and highlighted considerations for evaluating convergence. She highlighted conceptual models that researchers can draw from to help discern the elements and emphasized the need to move beyond publication metrics. In addition to the conceptual models from which researchers can draw to help discern the elements, it is important to establish structures that enable people to work together in a team or system and to design and develop processes and platforms to help teams create products that align with the team or system's vision and goals. A typical large research initiative at the National Institutes of Health (NIH), for example, might involve a governing body, research teams, working groups, and data centers, all of which must work together to create successful outcomes. She also distinguished the need for high-quality convergence and the requisite resources to support the capacity to conduct, manage, and support convergence research in terms of researchers to engage in the research, teams to manage the processes, and institutions with proper incentives and expectations.

Hall also considered what a portfolio of research might look like at the national level. How much convergence is desired? What mix of approaches beyond convergence might be considered? What are the dimensions that create different variations or types of convergence? What are the implications of these variations? Nested, interrelated levels of integration across departments and institutions, across projects and initiatives, and within projects are needed to support the readiness to undertake interdisciplinary and convergent projects, and this can prove self-reinforcing. Supportive university structures such as interdisciplinary centers can increase its likelihood of driving more interdisciplinary and convergent projects when designed to address barriers (distribution of indirect funds, for instance) and encourage facilitators (interaction across disciplines around common problems). She pointed to convergence profiles that differentiate by number of members (2, 10, 100, the latter possibly including teams of teams), number of disciplines, the distance between disciplines, the type of contributors (scholars, professionals, public, and patients), and the degree of integration (multidisciplinary, interdisciplinary, transdisciplinary), which highlights the heterogeneity of convergence collaborations. Likewise, she identified the need to consider variations in the types of convergence approaches such as research that brings existing datasets together, which one could label *post facto* convergence. She offered other examples, including *in vivo* convergence (dynamic, interactive idea generation), fundamental convergence, translational convergence, and social and intellectual convergence. These variations point to the need for research and evaluation to capture the depth and breadth of convergence and highlights how there is not a one-size-fits-all approach to supporting convergence.

Funding agencies can intervene to help facilitate the conduct of team research needed to pursue convergent science projects that is not happening organically due to existing barriers to collaboration. One such strategy includes issuing Funding Opportunity Announcements to help incentivize institutions to participate. By adding criteria that require grantees to collaborate across departments and institutions, funding agencies can create an imperative for institutional processes to follow. Paradoxically and with few exceptions, current tenure and promotion policies at U.S. institutions emphasize autonomy and independence and do not necessarily reward or recognize the kind of increasing interdependency needed for the complex topic of convergence. Initial descriptive and diagnostic methods can be used to understand what will enable convergent research. Given the discussions throughout the workshop that suggest we currently do not have the assessment tools needed to immediately launch a brief national survey, Hall concluded that at this juncture, our scientific enterprise may most need to recognize whether they have a context that enables convergence research so that it can happen when it is needed.

Holbrook addressed the value in constructing a national measure of convergence. Even if it is possible to do and could be done well, it does not automatically qualify as a wise decision if there are unintended consequences. It might end up harming people, fields of research, personnel within the research workforce, and communities impacted by the research. Holbrook urged NCSES to be very clear and upfront about stating its values: what it is trying to do, why, who it is intended to help, and who it is trying to avoid harming. He then turned to his second point, which emphasized making the initial effort: "to try something." He supported use of an iterative experimental approach that helps to incorporate feedback and ultimately reaches the intended goals without hurting fields of research, or people who are not directly involved in research, but are part of communities that are impacted by it. Though considerations about potential harm are valid, Holbrook favored action over fear of unintended consequences.

Khargonekar started to think about convergence soon after he joined the NSF as the assistant director for its Directorate of Engineering in

March 2013, a position he held until June 2016. He observed that the workshop led to a deeper and nuanced discussion of convergence. Khargonekar reminded the audience about the 2014 National Academies of Sciences, Engineering, and Medicine report on the deep integration of knowledge, tools, techniques, and ways of thinking from multiple disciplines as a novel method for producing new knowledge that is particularly suited to solving complex problems of societal interest.[2] He agreed with Ben Jones (Northwestern University) that an immense demand exists for such knowledge, but the demand encounters a bottleneck on the supply side, because the numbers of possible combinations are exponentially large and a priori knowledge of what would be truly successful is not available. Consequently, extremely large investments of resources are required. Considering convergence as a multidimensional and multifaceted concept and phenomenon, Khargonekar found the ecological framework proposed by Dan Stokols (University of California, Irvine) very compelling. Given this ecological view, all stakeholders have important roles in the convergence model of knowledge creation and their actions could help facilitate progress on measurements of convergence. The presenters' ideas ranged from the level of granularity at which such measurements might be made to various types of tagging, labeling, analysis, and a full life cycle of research from idea sharing to proposals to projects to products to impacts of research.

Khargonekar contended that a single measure of convergence at the national level is neither likely nor useful. He suggested looking for multiple indicators, maps, and other depictions of convergence. Quantitative and qualitative aspects of convergence must remain top of mind. There are already tools in various stages of development and maturity that hold great potential for mapping convergence. These tools should continue to be developed and made accessible to the research community and other stakeholders. As the entire research community becomes more familiar with these tools, it will be able to use them to improve all aspects of research processes and enhance the usefulness of the survey-based tools that NCSES uses.

If a problem-solving framework was accepted for convergence, then Khargonekar believed it would be logical for mission-driven agencies, such as the Environmental Protection Agency, the National Aeronautics and Space Administration, the National Oceanic and Atmospheric Administration, NIH, and the U.S. Departments of Defense and Energy, and others to take on the responsibility and allocate resources to address key problems of societal interest. He proposed that NSF take a leadership role in advancing

[2]National Research Council. (2014). *Convergence: Facilitating Transdisciplinary Integration of Life Sciences, Physical Sciences, Engineering, and Beyond*. Washington, DC: The National Academies Press.

the use of convergence research frameworks and measurement tools at mission agencies. As a pilot program, perhaps some key problem areas could be identified where these tools could be put to use to see if they shed new light on trajectories of progress.

Khargonekar considered it critical to have better tools for measuring convergence among doctoral graduates. A key challenge is assessing convergence within a doctoral dissertation that is meant to evaluate the work of one individual. Would the extent of convergence be determined by the publications that emerge from the dissertation or potentially from the diversity of the dissertation committee? How should the intellectual, emotional, and social components critical to convergence research in relation to the preparation and training of the PhD graduates be assessed? Would it be useful for NCSES to work with the Graduate Education Division at NSF on a national dialogue on doctoral student preparation, training, and dissertation production to gain insight on indicators? Could NCSES partner with other directorates and divisions at NSF that might be interested in measuring convergence among their research programs and funded researchers? How might NSF and NCSES work with industry on the majors and maps of convergence? Could convergence measurements and maps be useful for public-private partnerships? Khargonekar urged NCSES to make its tools more ubiquitous in the NCSES reports and publications both in print and online. He considered the portfolio-level maps to be most compelling and interesting.

Klein focused her comments on two topics: the value of constructing national-level measures and an overarching question of definition. Regarding the first point, she observed that presenters repeatedly raised lack of clarity and precision concerning how institutions and different researchers use multiple connotations of convergence as well as concepts of multidisciplinarity, transdisciplinarity, and interdisciplinarity. She emphasized that each term's definition is already grounded in authoritative literature, and consequently, efforts can now be made to create consistent language across institutions. These efforts will lead to rigor in the way that we approach integrative and collaborative research. Klein also highlighted a few caveats raised during the workshop. Nora Cate Schaeffer (University of Wisconsin–Madison) questioned, for example, whether a single definition is necessary, and asked how it might overlap with relevant concepts such as transdisciplinary synthesis. Graham Kalton (Westat) cautioned that we also need to ask who responds to surveys, as different people fill out a survey at different points.

To address her second point about definition, Klein stated that five keywords came to mind: purpose, focus, definition, unit, and methods. She observed that presenters throughout the workshop had challenged participants to clearly articulate the purpose of their efforts: is it monitoring or evaluation, progress or prediction, indicators or impacts? Is the goal to create a culture of convergence or to create a new field? Regarding focus: If a problem orientation is used, the audience and stakeholders need to be identified. As for definitions: Participants heard repeatedly that there are different types, discussed whether a more nuanced measure is also needed, and asked whether emerging disciplines could be factored in. For unit of analysis: The level of granularity reflects the relationship between simplicity and complexity. Levels might require different disciplines and criteria, or distinguish between project, goals, and products, or research versus policy, or even researchers versus commissions. Finally, what choices will be made about methods? Options might reflect a multidimensional approach, over-reliance on bibliometric data, the directionality of measures, or a portfolio approach.

Klein also addressed the steering committee's final question, which asked what NCSES might prioritize as part of its work in convergence. She recognized that the committee had not been tasked with providing formal recommendations, but she highlighted possible priorities that emerged from presenters. She referenced Kalton's pessimism about operating on a national scale at this time, Schaeffer's suggestion that convergence could expand over time, Khargonekar's caveat about not adopting one criterion, Owen-Smith's caveat regarding survey research limitations given ambiguities inherent in surveys with human subjects, Holbrook's concerns regarding the impact that standardization could have on convergence approaches, and Joseph DeSimone's (Stanford University) direct suggestion that trends must be considered. She then addressed the second question about ideas and strategies most worth pursuing, citing authoritative reports, white papers, and exemplar pilot studies consistent with an experimental mode. In considering possible products, Klein also noted a diversity of possible products ranging from surveys to topic modeling, interviews, maps, and data visualizations.

Owen-Smith began his remarks by positing that convergence is not a noun: It is not something that can exist or not exist or exist to some degree. It is a verb, a process that plays out across multiple levels of analysis that he as a sociologist believes are interrelated and causally affect each other. Convergence occurs over different timescales: a very micro timescale involving one-on-one in-person interactions among different people who may or may not feel particular affinities or frustrations, a medium term that is about the life span of a project, and a longer term that relates to fields of science, institutions, and careers. Owen-Smith encouraged the group to consider how the process could be unpacked over time. To the extent that researchers imagine a single cross-section of time, it makes sense to refer to convergence as a noun.

For those interested in being able to intervene, timing and temporal order are of significant interest to the convergence process. Treating convergence as a process highlights the need to consider who needs to be protected and empowered as efforts move forward. Acknowledgment, identification, and communication with stakeholders are essential. Owen-Smith also highlighted convergence's inherent dilemmas, particularly the potential trade-off between researcher visibility and productivity, which suggests there might be consequences particularly for scholars at earlier stages of their career and those less well connected to the core of their field. Convergence also creates a tension between the collective and the individual such that having a diverse team is advantageous for convergence but being the person who makes the team diverse and interesting can be individually disadvantageous. He observed that most of the socioemotional and organizational work required of interdisciplinary collaborations tends to fall to younger, particularly female scholars. Another dilemma relates to creative destruction, which emphasizes the creation but minimizes the destruction. This type of tension is also reflected in the need for strong disciplinary research, and how particular types of convergent research are supported.

In considering the convergence process, Owen-Smith favored a portfolio approach. Capturing convergence characteristics as they evolve will be important for understanding whether activities can accurately be considered convergent. He saw no reason, for example, why a convergent research process could not produce a disciplinary contribution. He added that to think about the characteristic of a particular entity, we need both point estimates and distributions. In the absence of interesting survey-based measures, he would recommend an iterative approach that holds multiple coordinated, concrete, community-engaged processes to capture particular aspects or measures in coordination with stakeholder groups. Owen-Smith contended that this approach might be particularly effective in arriving at a well-defined and well-articulated measure, particularly with input from an expert group to help avoid the drawbacks.

Smyth concurred with Owen-Smith's characterization of convergence as a process that crosses through time. This conceptualization has important implications for how we think about measuring convergence at a point in time, in particular the parts of the process we are not measuring and where errors tend to get introduced.

Conversations about interdisciplinarity and convergence research centers and their funding levels signify that a level of progress has already occurred. Some process must have brought the researchers together, and initializing factors must have led to convergence happening on a minute scale. Smyth raised a few related questions: First, at what point do participating individuals understand something as convergent and can report it as such, rather than underreporting convergent research that may be happening because they do not recognize it? Second, do errors related to the process and measures of convergence vary across organizations? And are some universities more likely to overlook convergence than others based on the metrics chosen and the point at which those metrics become important in the process?

As a survey methodologist, Smyth tends to break down components into error sources where the biases and variance affect the measure. She asked how we might choose a metric that accurately captures the process of convergence. Smyth agreed with Andrew Zukerberg's (National Center for Education Statistics) points relating to the use of definitions and their accompanying difficulties. Placing a shorter definition in a survey while providing a longer, more involved definition behind a hyperlink is unlikely to be the best method for standardizing this type of measurement. She noted that not all NCSES surveys are web-based: Paper questionnaires do still exist. More thought needs to go into what the measurement instruments are and how strategies will differ depending on varied survey modes.

Smyth returned to her starting point regarding opportunity costs. There are many ways to conduct this research, and each aspect does not need to be covered entirely by NCSES: Other agencies, organizations, and partnerships could form a team science approach with NCSES to address a robust measurement. Such an approach might maximize time-limited resources as well as limited surveys and respondents if researchers can be thoughtful about measurement and all of their priorities.

Entwisle began her closing remarks by identifying three components in the definition of convergence: (1) deep integration of multiple fields; (2) drawing on diverse viewpoints and experiences; and (3) solving "wicked" problems. Two of these components are part of the NSF definition, and one is not. She then addressed each of them, in turn.

With respect to the first component, Entwisle distinguished between deep integration and multiple fields and discussed them separately, beginning with the latter. In her view, an interest in the integration of multiple fields derives from a theory of progress based on crossing boundaries, where boundaries can be disciplines, departments, universities, and other entities, such as nations. She particularly wanted to draw attention to nations because the workshop did not include extensive discussion of international research. She noted that different ideas come from people with different experiences, from different places, and different points of view.

A challenge for measurement is that boundaries themselves may change. When convergence was introduced as a goal for research in 2010, the relevant fields were engineering, physical sciences, and biology. Since then, the list of relevant fields has expanded, the definition of fields has changed, and the boundaries themselves are changing. Entwisle said that typically, we do not want to change the measure if what we want to do is measure change. Posing the question of whether this was a problem, she responded, "maybe, or maybe that is the point." She emphasized that it is important to understand that the context of research is itself changing. Trying to standardize measurement without acknowledging that change is occurring can be counterproductive.

Entwisle then turned to some specific measures of boundary crossing. In her previous role as vice chancellor, she developed measures of convergence based on collaborations on funded research involving faculty from different departments and from different schools at the University of North Carolina at Chapel Hill. These metrics, based on numbers of and funds involved in collaborative projects, were relatively simple to construct and track. However, the metrics she developed were specific to the organization of her university and might look quite different at other academic institutions, such as Arizona State University. The measures Entwisle used did not match the information currently requested in the Higher Education Research and Development (HERD) survey either, although she saw potential for appropriate questions to be designed.

Entwisle contended that crossing of boundaries is a necessary condition for convergence, and relatively straightforward to measure. Deep integration is much more difficult to define and it is also more difficult to achieve. It may involve the social and emotional aspects of collaboration as well as the scientific ones. She said that deep integration is the difference between scientists each writing chapters for a book that will be stapled together versus being deeply engaged in creating an entirely new product. Due to the complexities, Entwisle expressed uncertainty about how well deep integration can be measured.

In Entwisle's view, the second component of convergence is diversity. This is not currently part of NSF's definition of convergence. She argued that truly convergent work requires more than linking disciplines and organizational units. Diverse viewpoints and experiences are also important. The Survey of Doctorate Recipients (SDR) is a potential source of this kind of information. The SDR collects personal information that can be linked with dissertation committees, dissertation abstracts, bibliometric databases, and the like. Similarly, the NCSES surveys could be linked with other administrative data collected by other federal government agencies or units.

She then addressed the third component of convergence: problem orientation. She agreed with others that convergence is needed to solve major societal problems, but she said that she was not convinced that solving a large-scale social problem needs to be a requirement for work to be considered convergent. She wondered how one would evaluate whether solving a major social problem is the motivation for a particular research project, given that scientists who are good at their craft are also good at explaining their research in this way. She then reflected on time frame: How long does it take to show progress on solving a challenging social problem, and how is success determined.

Entwisle said that the COVID-19 pandemic is a great example of a social problem, though it is a unique case because much of the world is trying to solve it. Many social problems will not receive that level of attention and widespread efforts. As an example of convergence research that receives less attention, Entwisle mentioned a project focused on storm inundation on the eastern North Carolina coast, which brings together scientists from several fields, including engineering, ecology, geography, and various social sciences to solve a problem.

Entwisle argued that some convergent research is not immediately applicable to a problem. She added that if application is to be a required component of convergence research, then additional reflection about businesses and other nonacademic units and their collaborations as well as translational research is needed. These were not discussed very much in the workshop.

Entwisle concluded by saying that while we do not have all of the answers, this should not stand in the way of progress. NSF needs information about convergence research to manage its own research portfolio and also to advise on the national portfolio.

In her closing remarks, Emilda Rivers expressed NCSES' interest in remaining connected with the community on best approaches to follow moving forward. She pointed out that NCSES is a federal statistical agency with a mission to report on the science and engineering enterprise in a frequently global context and receives several requests for new areas and topics to measure. This workshop allowed participants to see the breadth of knowledge that NCSES covers from both the national and international science and engineering communities.

Rivers said that she found ideas for leveraging administrative records to assist NCSES efforts especially interesting because of concerns associated with the respondent burden posed by survey data collections. Another concern that resonated with Rivers is the need for measures to make sense to a respondent. She encouraged participants and attendees as members of the research community to sustain and remain vigilant on the status of convergence and to continue thinking about aspects that should be studied and insightful avenues that would move progress forward.

Appendix A

Workshop Agenda

Workshop on the Implications of Convergence for How the National Center for Science and Engineering Statistics Measures the Science and Engineering Workforce

October 22–23, 2020

THURSDAY, OCTOBER 22

10:00–10:15 a.m.
Welcome and plans for the workshop

BRIAN HARRIS-KOJETIN, Director, Committee on National Statistics
BARBARA ENTWISLE, Steering Committee Chair, University of North Carolina at Chapel Hill

10:15–10:30 a.m.
The National Science Foundation's interest in measuring convergence

ARTHUR LUPIA, Assistant Director, National Science Foundation Directorate for Social, Behavioral, and Economic Sciences

10:30–10:45 a.m.
The National Center for Science and Engineering Statistics' goals for the workshop

EMILDA RIVERS, Director, National Center for Science and Engineering Statistics

10:45–11:10 a.m.
Setting the stage: Why is convergence important to measure?

JOSEPH DESIMONE, Stanford University and Carbon, Inc.

11:10–11:25 a.m.
Break

11:25–12:10 p.m.
Overview of NCSES' data collections, data products, and previous work in this area

GARY ANDERSON, Senior Science Resource Analyst, National Center for Science and Engineering Statistics

12:10–1:00 p.m.
Break

1:00–1:30 p.m.
Definitions of convergence

Moderator: JULIE THOMPSON KLEIN, Wayne State University (emerita) and Transdisciplinarity Lab at ETH-Zurich

DAN STOKOLS, University of California, Irvine

1:30–2:50 p.m.
Convergence as practiced

Moderators: PRAMOD KHARGONEKAR, University of California, Irvine and JAMES BRITT HOLBROOK, New Jersey Institute of Technology

LORA WEISS, Pennsylvania State University

PETER SCHIFFER, Yale University

DAN GALLAHAN, National Cancer Institute

MARYANN FELDMAN, University of North Carolina at Chapel Hill

2:50–3:00 p.m.
Break

3:00–4:00 p.m.
Outcomes and impacts of convergence

Moderators: KARA HALL, National Cancer Institute, and JASON OWEN-SMITH, University of Michigan

JULIA LANE, New York University

ERIN LEAHEY, University of Arizona

BEN JONES, Northwestern University

FRIDAY, OCTOBER 23

10:00–10:10 a.m.
Welcome and overview of Day 2

BARBARA ENTWISLE, Steering Committee Chair, University of North Carolina at Chapel Hill

10:10–11:15 a.m.
Measurement efforts

Moderators: JOLENE SMYTH, University of Nebraska and KARA HALL, National Cancer Institute

BETHANY LAURSEN, Michigan State University
(with Nicole Motzer, National Socio-Environmental Synthesis Center and Kelly Anderson, University of Maryland)

MICHÈLE LAMONT, Harvard University

DANIEL MCFARLAND, Stanford University

11:15–11:30 a.m.
Break

11:30 a.m.–12:15 p.m.
Measurement efforts (continued)

Moderators: James Britt Holbrook, New Jersey Institute of Technology, and Pramod Khargonekar, University of California, Irvine

Ismael Rafols, University of Leiden, Netherlands

Joshua Schnell, Clarivate

12:15–1:15 p.m.
Break

1:15–2:15 p.m.
Panel discussion on implementation and feasibility

Moderators: Jason Owen-Smith, University of Michigan and Jolene Smyth, University of Nebraska

Graham Kalton, Westat

Nora Cate Schaeffer, University of Wisconsin–Madison

Andrew Zukerberg, National Center for Education Statistics

2:15–3:30 p.m.
Summary of key themes and thoughts on how to move this work forward

Steering committee members will reflect on key takeaways from the workshop, including:

- What would be the value in constructing national-level measures of convergence?
- What ideas and strategies would be most worth pursuing to push thinking on this initiative forward?
- What framework could NCSES use for measuring convergence?
- What definitions and criteria would we consider appropriate and what are their limitations?
- What types of NCSES products would be most valuable and for what audiences?
- What should NCSES prioritize as part of its work in this area?

KARA HALL, National Cancer Institute
JAMES BRITT HOLBROOK, New Jersey Institute of Technology
PRAMOD P. KHARGONEKAR, University of California, Irvine
JULIE THOMPSON KLEIN, Wayne State University (emerita) and Transdisciplinarity Lab at ETH-Zurich
JASON OWEN-SMITH, University of Michigan
JOLENE SMYTH, University of Nebraska
BARBARA ENTWISLE, University of North Carolina at Chapel Hill

Appendix B

Biographical Sketches of the Steering Committee Members and Workshop Speakers

GARY ANDERSON is a senior economist in the Research and Development Statistics Program at the National Center for Science and Engineering Statistics (NCSES) within the National Science Foundation (NSF). At NCSES, he works with business R&D and innovation data. Dr. Anderson works extensively with the Business Enterprise Research and Development Survey, Annual Business Survey, and, along with a team of researchers, conducts research on the measurement of business innovation using machine learning and other data science techniques. Much of his work focuses on novel uses of business microdata and has contributed to international efforts to explore the structure, distribution, and concentration of business R&D. Prior to joining NCSES, Anderson was a senior economist in the Director's Office at the National institute of Standards and Technology (NIST). Anderson formulated and led strategic planning and economic impact analysis on standards, technology infrastructure, and technology transfer. He also served as senior economist with NIST's Advanced Technology Program. Anderson holds a PhD in economics from the University of Maryland.

JOSEPH DESIMONE is the Sanjiv Sam Gambhir Professor of Translational Medicine and Chemical Engineering in the Department of Radiology, the Department of Chemical Engineering, and the Graduate School of Business (by courtesy) at Stanford University. He is cofounder and board chair of Carbon, an innovative 3D printing company valued at almost $2.5 billion. He is also Chancellor's Eminent Professor of Chemistry Emeritus at the University of North Carolina at Chapel Hill and the William R. Kenan,

Jr. Distinguished Professor of Chemical Engineering Emeritus at North Carolina State University. He made scientific breakthroughs in areas including green chemistry, medical devices, and nanotechnology. He also cofounded several companies based on his research. He is the author of over 350 scientific publications and a named inventor on over 200 issued patents. In recognition of his entrepreneurial success at Carbon, Dr. DeSimone was recently recognized as the 2019 EY Entrepreneur of the Year U.S. National Overall winner. He was a recipient of the National Medal of Technology and Innovation from President Obama. He is one of only 25 individuals elected to all three U.S. National Academies (Sciences, Medicine, and Engineering) and a former chair of the National Academies Committee on Key Challenge Areas for Convergence and Health. Dr. DeSimone has a PhD in chemistry from Virginia Polytechnic Institute and State University.

BARBARA ENTWISLE (*steering committee chair*) is Kenan Distinguished Professor of Sociology and fellow of the Carolina Population Center (CPC) at the University of North Carolina (UNC) at Chapel Hill. She also holds adjunct and affiliate faculty positions in geography and the environment, energy, and ecology. Prior to this, Entwisle served as vice chancellor for research at UNC at Chapel Hill and led a campuswide research program with nearly $1 billion in annual research expenditures. Dr. Entwisle also oversaw 10 research development, support, and compliance offices and managed 15 pan-university research centers and institutes, including the CPC. Her research efforts focus on two main areas—population/environment interactions and migration and the life course. Entwisle has a PhD in sociology from Brown University.

MARYANN FELDMAN is the Heninger Distinguished Professor in the Department of Public Policy at the University of North Carolina, an adjunct professor of finance at Kenan-Flagler Business School, and a research director at UNC Kenan Institute of Private Enterprise. From 2014–2017, Dr. Feldman held a joint appointment at the National Science Foundation as the Science of Science and Innovation Policy (SciSIP) Program Director and chaired an interagency working group on Science Policy. Her research and teaching interests focus on the areas of innovation, the commercialization of academic research, and the factors that promote technological change and economic growth. Dr. Feldman was the winner of the 2013 Global Award for Entrepreneurship Research for her contributions to the study of the geography of innovation and the role of entrepreneurial activity in the formation of regional industry clusters. Currently, Feldman is researching the industrial genesis of the Research Triangle region. The project follows the development of the regional economy over a 50-year time period using a unique database of 3,200 entrepreneurial ventures and attempts to understand the institutional dynamics that created a vibrant regional economy. She has a PhD in economics and management from Carnegie Mellon University.

DAN GALLAHAN is the director of the Division of Cancer Biology at NIH's National Cancer Institute (NCI). He established the division's efforts in cancer systems biology, the latest of which is the Cancer Systems Biology Consortium. Prior to that, he also had an NCI intramural career at the Laboratory of Tumor Immunology and Biology. His experience outside the government includes serving as director of molecular biology and development for a small biotech company. He is a molecular and cancer biologist with expertise in the fields of breast cancer, technology development, and science administration. His primary focus is the integration of multiple approaches, tools, and datasets to understanding the complexity of cancer. Gallahan received his doctoral training in molecular biology and biochemistry from the University of Maryland.

KARA HALL (*steering committee member*) is a health scientist, director of the Science of Team Science (SciTS), and program director at the National Cancer Institute. She helped launch the SciTS field through her leadership in conducting empirical studies, developing conceptual frameworks, creating evidence-based strategies and resources, editing special journal issues, chairing the Annual SciTS Conferences, and serving as founding board member of International Network for the SciTS. Dr. Hall has led efforts to support team science across funding agencies in the United States and internationally such as a transagency subcommittee on collaboration and team science of the Federal Networking and Information Technology Research and Development Program in the Executive Office of the President's National Science and Technology Council, on which she served as cochair. She is contributing editor of the book *Strategies for Team Science Success: Handbook of Evidence-Based Principles for Cross-Disciplinary Science and Practical Lessons Learned from Health Researchers*, which has been downloaded more than 32,000 times. During her career, Dr. Hall has contributed to advancing health behavior research, implementation science, systems science approaches, and research on sleep and cancer. She received her PhD in psychology from the University of Rhode Island.

BRIAN HARRIS-KOJETIN is the director of the Committee of National Statistics (CNSTAT). He joined CNSTAT as deputy director in 2015 from the Office of Management and Budget (OMB), where he served as a senior statistician in the Statistical and Science Policy Office. He chaired the Federal Committee on Statistical Methodology and was the lead at OMB on issues related to standards for statistical surveys, survey nonresponse, survey respondent incentives, and the Confidential Information Protection and Statistical Efficiency Act of 2002. He also served as the desk officer for the Bureau of Labor Statistics and the demographic programs of the U.S. Census Bureau. Prior to joining OMB in 2001, he was the senior project leader for research standards and practices at the Arbitron Company. He also previously served as a research psychologist in the Office of Survey Methods Research at the Bureau of Labor Statistics. He is a fellow of the American Statistical Association (ASA). Harris-Kojetin has a PhD in social psychology with a minor in statistics from the University of Minnesota.

JAMES BRITT HOLBROOK (*steering committee member*) is an associate professor in the Department of Humanities at the New Jersey Institute of Technology. Holbrook has held teaching positions at Emory University, Georgia State University, and Georgia Tech and was previously research assistant professor in the Department of Philosophy and Religion Studies at the University of North Texas, where he also served as assistant director of the Center for the Study of Interdisciplinarity. His teaching and research explore the value of knowledge and his work focuses on knowledge institutions (the university and science and technology funding agencies, in particular), the evaluation of knowledge (involving the inclusion of scholarly outputs and broader societal impacts), and the role knowledge users ought to play in answering the question of the value of knowledge (moving beyond coproduction of knowledge to explore the question of who should count as a peer). Dr. Holbrook has a PhD in philosophy from Emory University.

BENJAMIN JONES is the Gordon and Llura Gund Family Professor of Entrepreneurship, a professor of strategy, and the faculty director of the Kellogg Innovation and Entrepreneurship Initiative. An economist by training, Jones studies the sources of economic growth in advanced economies with an emphasis on innovation, entrepreneurship, and scientific progress. He also studies global economic development, especially the roles of education, climate, and national leadership in explaining the wealth and poverty of nations. His research has appeared in journals such as *Science*, the *Quarterly Journal of Economics*, and the *American Economic Review*. He has also been profiled in media outlets such as the *Wall Street Journal*, the *Economist*, and *The New Yorker*. A former Rhodes Scholar, Dr. Jones served in 2010–2011 as the senior economist for macroeconomics for the White House Council of Economic Advisers and earlier served in the U.S. Department of the Treasury. Jones is a nonresident senior fellow of the Brookings Institution, a research associate of the National Bureau of Economic Research, and a member of the Council on Foreign Relations. He has a PhD in economics from the Massachusetts Institute of Technology.

GRAHAM KALTON is a nonresident senior statistical fellow and former chairman of the board of directors of Westat. He is also a research professor in the Joint Program in Survey Methodology at the University of Maryland. He is an internationally known survey statistician and methodologist who has conducted research on many aspects of survey methodology and has been a consultant on a number of federal statistical programs. Dr. Kalton is a fellow of the American Statistical Association, a fellow of the American Association for the Advancement of Science, an elected member of the International Statistical Institute, and a national associate of the National Academy of Sciences. He is a recipient of the Jerzy Neyman medal from the Polish Statistical Association, the Morris Hansen Lecture Award, and the Waksberg Award. Dr. Kalton has served as a member and chair of statistics of Canada's Advisory Committee on Statistical Methods. He also served as a member of the Committee for National Statistics of the National Academies, the Board of Scientific Counselors for the National Center for Health Statistics, and the Federal Economic Statistics Advisory Committee. He has a PhD in survey statistics from the University of Southampton, United Kingdom.

PRAMOD P. KHARGONEKAR (*steering committee member*) is vice chancellor for research and Distinguished Professor of Electrical Engineering and Computer Science at the University of California, Irvine. He is an expert in control and systems theory, as well as applications to manufacturing, renewable energy, and biomedical engineering. Dr. Khargonekar is particularly interested in understanding and improving processes that connect research to innovation. He has served on the faculty at the University of Florida, University of Minnesota, University of Michigan, and University of California, Irvine. In 2012–2013, he served as deputy director of technology at the U.S. Department of Energy's Advanced Research Projects Agency-Energy (ARPA-E), and in 2013, the National Science Foundation (NSF) appointed him to serve as assistant director for the Directorate of Engineering, a position he held until June 2016. At NSF, Dr. Khargonekar promoted efforts for multidisciplinary collaborations to address societal grand challenges. He also contributed to several of NSF's Big Ideas, as well as science, technology, engineering, and mathematics issues at all educational levels with an emphasis on improving opportunities and outcomes for women and underrepresented minorities in engineering. He has a PhD in electrical engineering from the University of Florida.

JULIE THOMPSON KLEIN (*steering committee member*) is professor of humanities (emerita) in the English Department and former faculty fellow for interdisciplinary development in the Division of Research at Wayne State University, as well as an international research affiliate in the Transdisciplinarity Lab at the ETH-Zurich's Swiss Federal Institute of Technology. Dr. Klein is a past president of the Association for Interdisciplinary Studies (AIS) and former editor of the AIS journal, *Issues in Integrative Studies*. She has authored several books including *Interdisciplinarity: History, Theory, and Practice*; *Crossing Boundaries: Knowledge, Disciplinarities, and Interdisciplinarities*; *Humanities, Culture, and Interdisciplinarity: the Changing American Academy*; and *Interdisciplining Digital Humanities*. She is about to release a new book, *Beyond Interdiciplinarity: Boundary, Communication, and Collaboration*. She has also co-edited several books including *Interdisciplinary Studies Today*; *Transdisciplinarity: Joint Problem Solving among Science, Technology, and Society;* and *The Oxford Handbook on Interdisciplinarity*. In addition to helping numerous colleges and universities develop interdisciplinary programs, Dr. Klein was a member of the Association of American Colleges and Universities' first national Task Force on Interdisciplinary Studies. She has a PhD in English from the University of Oregon.

MICHÈLE LAMONT is professor of sociology and of African and African American studies and the Robert I. Goldman Professor of European Studies at Harvard University. She is also the director of the Weatherhead Center for International Affairs at Harvard University. A cultural and comparative sociologist, Dr. Lamont is the author or coauthor of a dozen books and edited volumes and more than 100 articles and chapters on a range of topics including culture and inequality, racism and stigma, academia and knowledge, social change and successful societies, and qualitative methods. Dr. Lamont served as the 108th president of the American Sociological Association in 2016–2017 and she chaired the Council for European Studies (2006–2009). She is also the recipient of a 1996 John Simon Guggenheim fellowship, the 2014 Gutenberg research award, and the 2017 Erasmus prize (for her contributions to the social sciences in Europe and the rest of the world). She is also the recipient of honorary doctorates from five countries (Canada, France, the Netherlands, Sweden, and the United Kingdom). An Andrew Carnegie fellow for 2019–2021, she spent 2019–2020 on sabbatical at the Russell Sage Foundation. Dr. Lamont has a PhD in sociology from the Université de Paris.

JULIA LANE is a professor at the New York University (NYU) Wagner Graduate School of Public Service and the NYU Center for Urban Science and Progress. She is also NYU Provostial Fellow for Innovation Analytics. Previous to this, Dr. Lane was a senior managing economist and institute fellow at the American Institutes for Research. She also held positions at the National Science Foundation, the Urban Institute, the World Bank, American University, and NORC at the University at Chicago. In these positions, Dr. Lane has led many initiatives, including cofounding the Institute for Research and Innovation in Science (IRIS) at the University of Michigan and the STAR METRICS programs at the National Science Foundation. She also initiated and led the creation and permanent establishment of the Longitudinal Employer-Household Dynamics Program at the U.S. Census Bureau. Dr. Lane has published over 80 articles in leading economics journals and authored or edited 10 books. She is an elected fellow of the American Association for the Advancement of Science and the International Statistical Institute, and a fellow of the American Statistical Association. She is the recipient of the 2014 Julius Shiskin award, the 2014 Roger Herriot award, and the 2017 Warren E. Miller award. She received her PhD in economics from the University of Missouri.

BETHANY LAURSEN is principal consultant at Laursen Evaluation & Design and assistant dean of the Michigan State University (MSU) graduate school. She is a scholar-practitioner of integration and implementation sciences (I2S), an emerging discipline that improves research impact on complex, socioenvironmental problems. After practicing as an interdisciplinary environmental scholar and educator for 10 years, she switched fields to study and facilitate interdisciplinarity itself. Her specialties are integrative and evaluative reasoning and tools that help engage in such reasoning more effectively. Drawing on her work as an experiential educator and facilitator, she is expanding her work into noncognitive ways of knowing. Dr. Laursen is a member of the Toolbox Dialogue Initiative, a research and facilitation project based at MSU that promotes communication in cross-disciplinary teams using philosophical dialogue. She has a PhD in community sustainability from Michigan State University.

ERIN LEAHEY is professor and director of sociology at the University of Arizona. Her main research interests are scientific practice, scientific careers, and innovation. She also researches inequality in science and scientific careers, particularly along gender lines. Recently, she has conducted investigations of knowledge that spans domains (both subfields and fields). She is developing measures of the commitment of universities to interdisciplinarity and the assessment of both the drivers of commitment levels and their consequences for scientific research output and university prestige. She is also examining whether and how large secondary datasets serve as hubs that help scholars from multiple disciplines coalesce. Dr. Leahey won an award from Cambridge Scientific Abstracts for creative use of its data to develop measures of two theoretical constructs relevant to scientific careers: the extent of specialization and subfield integration. She has a PhD in sociology from the University of North Carolina at Chapel Hill.

ARTHUR LUPIA heads the Directorate for Social, Behavioral, and Economic Sciences at the National Science Foundation. He is also the Gerald R. Ford Distinguished University Professor and Hal R. Varian Collegiate Professor of Political Science at the University of Michigan. He has held a range of scientific leadership positions including principal investigator of the American National Election Studies and has worked with many groups to improve decision-making and the communication of scientific facts. He has served on the boards of organizations dedicated to increasing the social value of scientific research including the Center for Open Science, the National Academies' Advisory Board on the Behavioral and Social Sciences and Education, and the American Association for the Advancement of Science's Leshner Leadership Institute. He has served as chair of the AAAS Section on Social, Economic, and Political Sciences, president of the Midwest Political Science Association, and a range of leadership positions at the American Political Science Association, including treasurer and chair of the Task Force on Public Engagement. Dr. Lupia received multiple honors including the Ithiel de Sola Pool Award from the American Political Science Association and the National Academy of Sciences' Award for Initiatives in Research. He is an elected member of the American Academy of Arts and Sciences, a fellow of the American Association for the Advancement of Science. He has also been a Guggenheim fellow and is one of the inaugural Andrew Carnegie fellows. Dr. Lupia has a PhD in social science from the California Institute of Technology.

DANIEL McFARLAND is professor of education and, by courtesy, of sociology and organizational behavior at the Graduate School of Business at Stanford University. He studies the social and organizational dynamics of educational systems like schools, classrooms, and universities. He conducted studies on classroom organization and interaction; the formation of adolescent relationships, social structures, and identities; interdisciplinary collaboration and intellectual innovation; and relational sociology. His broad research interests resulted in a variety of interdisciplinary collaborations with linguists and computer scientists. This in turn has led to studies of big data and methodological advances in social networks and language modeling. He is the recipient of the Gould award from the *American Journal of Sociology* (2013–2014) and of the Friedrich Wilhelm Bessel award from the Alexander von Humboldt Foundation (2014–2017). McFarland has a PhD in sociology from the University of Chicago.

JASON OWEN-SMITH (*steering committee member*) is a research professor in the Institute for Social Research Survey Research Center and professor of sociology at the University of Michigan. He is also a cofounder and the executive director of the Institute for Research on Innovation and

Science, which collects, curates, and shares data to help understand, explain, and eventually improve the public value of investments in academic research. Dr. Owen-Smith works on projects that examine the dynamics of high-technology industries, the public value of the research university, and the network organization of surgical care. He seeks to understand how organizations, institutions, and networks can maintain the status quo while generating novelty through social transformations, scientific discoveries, and technological breakthroughs. He is the author of the book *Research Universities and the Public Good: Discovery for an Uncertain Future*. Dr. Owen-Smith has a PhD in sociology from the University of Arizona.

ISMAEL RAFOLS is a senior researcher at the Center for Science and Technology Studies (CWTS) at Leiden University in the Netherlands, and an associate faculty member in the Science Policy Research Unit at the University of Sussex. He is also a member of the Science, Technology & Innovation Studies Group. Before CWTS, he worked at the University of Sussex and INGENIO, a joint research center of the Spanish National Research Council and the Universitat Politècnica de València. In between academic positions, he has worked on international cooperation in Oxfam and the City Council of Barcelona. He currently works on science policy, developing novel approaches to S&T indicators, and using mixed methods for informing evaluation, foresight, and research strategies. Dr. Rafols has been involved in initiatives on "responsible metrics" such as the Leiden Manifesto, the EC Expert Group on Open Science Indicators, which involved discussions on biases against research in "peripheral" topics and regions. Previously, he had developed indicators and mapping methods for the evaluation of interdisciplinary research in such emergent fields as biotech and nanotechnology. He received a PhD in biophysics from Tohoku University in Japan.

EMILDA RIVERS is the director of the National Center for Science and Engineering Statistics (NCSES). Prior to her appointment as NCSES director, she was the NCSES deputy director. She previously led the center's largest program area, the Human Resources Statistics Program. She has extensive experience in the federal statistical service as a mathematical statistician working on establishment and demographic surveys, including at the U.S. Census Bureau and U.S. Energy Information Administration. During these years of service, she has often used her statistical and methodological expertise to design and implement projects that provide previously missing data critical to decision making. In 2017, she was named by Forbes as one of 25 Women Leading Data and Analytics in the U.S. government. She has an MS degree from the University of Maryland, College Park.

NORA CATE SCHAEFFER is Sewell Bascom Professor of Sociology (emerita) at the University of Wisconsin–Madison. She is former faculty director of the Survey Center and interim associate vice chancellor for research in the social sciences at the University of Wisconsin–Madison. She has also taught at the Summer Institute of the Survey Research Center at the University of Michigan and through the University of Michigan-University of Maryland Joint Program in Survey Methodology. Her teaching and research focuses on survey research methods, particularly instrument design and survey measurement and on interaction in the survey interview. Schaeffer was elected to a 3-year presidential term (vice president, president, and past president) of the American Association for Public Opinion Research beginning in May 2018. She is a fellow of the American Statistical Association and a fellow of the Midwest Association for Public Opinion Research. She has served on numerous advisory committees and boards that advise federal agencies on data collection methods. She has a PhD in sociology from the University of Chicago.

PETER SCHIFFER is the Frederick W. Beinecke Professor of Applied Physics at Yale University and senior fellow at the Association of American Universities. He also served as the inaugural vice provost for research at Yale University. Dr. Schiffer is an international leader in experimental condensed matter physics, specializing in the study of magnetic systems. Prior to joining Yale University, he served as vice chancellor for research at the University of Illinois at Urbana-Champaign and as associate vice president for research and director of strategic initiatives at Pennsylvania State University. He undertook postdoctoral work at AT&T Bell Laboratories before launching his faculty career at the University of Notre Dame. He is a fellow of the American Physical Society and the recipient of academic honors including an Alfred P. Sloan Research fellowship, a Presidential Early Career award from the Army Research Office, and a National Science Foundation Early Career Development award. He has a PhD in physics from Stanford University.

JOSHUA SCHNELL is a senior advisor at the Institute for Scientific Information within the Web of Science Group at Clarivate Analytics, and he advises on science planning, the evaluation of R&D, and science and technology policy. Prior to this, he oversaw a team responsible for conducting program evaluations of research and training programs and developing technology for data-driven science management. Before joining ISI, he directed the analytics group at a science management startup, worked in research administration at Northwestern University, and was a science and technology policy graduate fellow at the National Academies of Sciences, Engineering, and Medicine. He holds a PhD from Northwestern University.

JOLENE SMYTH (*steering committee member*) is associate professor and director of the Bureau of Sociological Research at the University of Nebraska-Lincoln. Her research focuses on survey methodology and gender. Smyth believes strongly in the importance of high quality data to improve science and policy, and the methodology research questions she pursues are largely shaped by changes occurring in the survey industry that raise important questions about data quality. Smyth's recent work focuses on improving the National Science Foundation's science and engineering surveys. She is coauthor of the book *Internet, Phone, Mail, and Mixed-Mode Surveys: the Tailored Design Method*, and she serves on the editorial board of *Public Opinion Quarterly*. In 2017, she was corecipient of the Warren J. Mitofsky Innovator's Award from the American Association for Public Opinion Research. She has a PhD in sociology from Washington State University.

DANIEL STOKOLS is Chancellor's Professor Emeritus in the School of Social Ecology at the University of California, Irvine (UCI). He served as founding dean of the School of Social Ecology. His research focuses on the study of people's transactions with their social and physical environments and spans the fields of social ecology, environmental psychology, urban planning, epidemiology, and public health. Dr. Stokols also does research on the science of team science and investigates factors that affect the collaborative success of transdisciplinary research and training programs. He currently codirects UCI's Team Science Acceleration Lab. Dr. Stokols is the recipient of the career award from the Environmental Design Research Association, UCI's Lauds & Laurels Faculty Achievement award, Newman-Proshansky Career award from the American Psychological Association's Division of Environmental, Population, and Conservation Psychology, the International Network for the Science of Team Science Career Recognition award, and the School of Social Ecology Founders award. Dr. Stokols has served as a team science evaluation consultant for the National Cancer Institute, the National Academies Keck Futures Initiative, and as a member of the National Academies of Sciences, Engineering, and Medicine Committee on the Science of Team Science. He has a PhD in social psychology from the University of North Carolina at Chapel Hill.

LORA WEISS is the senior vice president for research at the Pennsylvania State University, where she oversees the research work of eleven academic colleges, seven university-wide interdisciplinary research institutes, a university-affiliated research center for the Department of Defense, and offices for sponsored programs, research protections, industry partnerships, technology transfer, innovation, and commercialization. Prior to her current position, Dr. Weiss spent 13 years at Georgia Tech, where most recently, she was the interim director of the Georgia Tech Research Institute. She also worked at the Penn State Applied Research Laboratory. Her research is in the areas of robotics and unmanned systems. She was on the technical advisory board of the National Robotics Technology Consortium and the executive board of the National Defense Industrial Association Undersea Warfare Division, and she supported the Center for New American Security's Future Foundry Task Force. Dr. Weiss was named Regents researcher by the University System of Georgia. She is the recipient of the AUVSI Foundation Award for Academic Champion, and Georgia's Women in Technology Woman of the Year Award for Medium-Sized Businesses, and a letter of commendation from the Chief of Naval Operations Strategic Studies Group. She received her PhD in acoustics from Penn State.

ANDREW ZUKERBERG is the chief of the Cross-Sectional Surveys Branch at the National Center for Education Statistics (NCES) within the U.S. Department of Education. Prior to joining NCES, he managed qualitative and quantitative research studies at Gallup. He also served as a branch chief at the U.S. Census Bureau, where he managed large-scale data collections. Previously, he conducted product development research at Microsoft, where the methods and results of one of his studies were highlighted as a company-wide best practice. Throughout his career, Zukerberg's research focus has been on reducing nonsampling errors in surveys and improving response rates. He has an MS in survey methodology from the Joint Program in Survey Methodology at the University of Maryland.

COMMITTEE ON NATIONAL STATISTICS

The Committee on National Statistics was established in 1972 at the National Academies of Sciences, Engineering, and Medicine to improve the statistical methods and information on which public policy decisions are based. The committee carries out studies, workshops, and other activities to foster better measures and fuller understanding of the economy, the environment, public health, crime, education, immigration, poverty, welfare, and other public policy issues. It also evaluates ongoing statistical programs and tracks the statistical policy and coordinating activities of the federal government, serving a unique role at the intersection of statistics and public policy. The committee's work is supported by a consortium of federal agencies through a National Science Foundation grant, a National Agricultural Statistics Service cooperative agreement, and several individual contracts.